Otherworldliness Depicted:
Measured Drawings for Classical Gardens by
Xi'an University of Architecture and Technology, 2011-2014

LIN Yuan, YUE Yanmin, WEN Wujuan, LIN Xi

尘外壶中
——西安建筑科技大学古典园林测绘图辑 2011—2014

林源　岳岩敏　汶武娟　林溪　著

中国建筑工业出版社

本书编写人员名单：

	林　源　　岳岩敏　　汶武娟　　林　溪
文　字	林　源　　岳岩敏　　林　溪
英　文	林　溪
版式设计	汶武娟
图纸绘制	汶武娟　　王茹悦　　李双双　　杨烜子　　王艺博　　张文波 陈斯亮　　马英晨　　魏　颀

Credits:

Major Contributors: LIN Yuan, YUE Yanmin, WEN Wujuan and LIN Xi

Text: LIN Yuan, YUE Yanmin and LIN Xi

English Translation: LIN Xi

Book Design: WEN Wujuan

Drawing Preparation: WEN Wujuan, WANG Ruyue, LI Shuangshuang, YANG Xuanzi, WANG Yibo, ZHANG Wenbo, CHEN Siliang, MA Yingchen and WEI Qi

序 Perface

中国古建筑测绘与研究的守望者

校园里的林源教授，清雅宁静，怡然自得，与她的学术追求互相映衬，相得益彰。

提及中国古建测绘，必然想起梁思成、林徽因等先辈从1931年起，在15年间踏遍中国十五省，对190个县2738处古建进行测绘和拍摄的创举；就会想起林徽因攀着木梯测绘唐代经幢的照片画面……

林源老师的"古建筑测绘之路"亦是如此：自1998年首开了我校的中国古建筑测绘课程，带着学生和图板，从陕西出发，走过甘肃、山西、西藏、辽宁……以地区为单位开展田野调查和测绘记录；八下江南，完成了九处江南古典园林的测绘工作，至今整整二十年了。她谨记导师赵立瀛教授嘱托：测绘是古建筑研究的木之根、水之源，于是她抱质守拙，坚守一线测绘教学与研究，亲手绘制了大量的第一手测绘图纸；她从不把"测绘"看作简单的复制，于是她的"测绘"是带着情感的工作，这种情感是对前人劳动的一种体会，是与古代工匠智慧的一场对话；她倾心于我校历史建筑保护工程专业教学体系的构建，守望着华夏大地众多建筑遗产保护工作的传承与延续，守望着"师"者的初心。

守望者并不孤单，越来越多的青年教师加入团队，这卷《尘外画中——西安建筑科技大学古典园林测绘图辑2011-2014》即是林源教授和"小伙伴"们一道工作所取得的部分成果。很高兴看到它们可以呈现在读者面前。

是以为序，并以此共同守望。

西安建筑科技大学校党委常委、建筑学院党委书记

2019年1月30日

目录

序

苏州艺圃 ·················001

苏州环秀山庄 ·················040

苏州耦园 ·················065

苏州怡园 ·················133

苏州沧浪亭 ·················169

后记 ·················246

Contents

Preface

Yi Pu (Garden of Cultivation), Suzhou 001

Huanxiu Shanzhuang (Mountain Villa with Embracing Beauty), Suzhou 040

Ou Yuan (Couple's Garden Retreat), Suzhou 065

Yi Yuan (Garden of Delight), Suzhou 133

Canglang Ting (Canglang Pavilion), Suzhou 169

Postscript 247

苏州艺圃－园池南岸立面图

Elevation: south bank of the Pond, Yi Pu, Suzhou

苏州环秀山庄 – 全园南北剖面（东视）
North-south section: Huanxiu Shanzhuang, Suzhou

苏州环秀山庄－园池南立面图

South elevation: The Pond, Huanxiu Shanzhuang, Suzhou

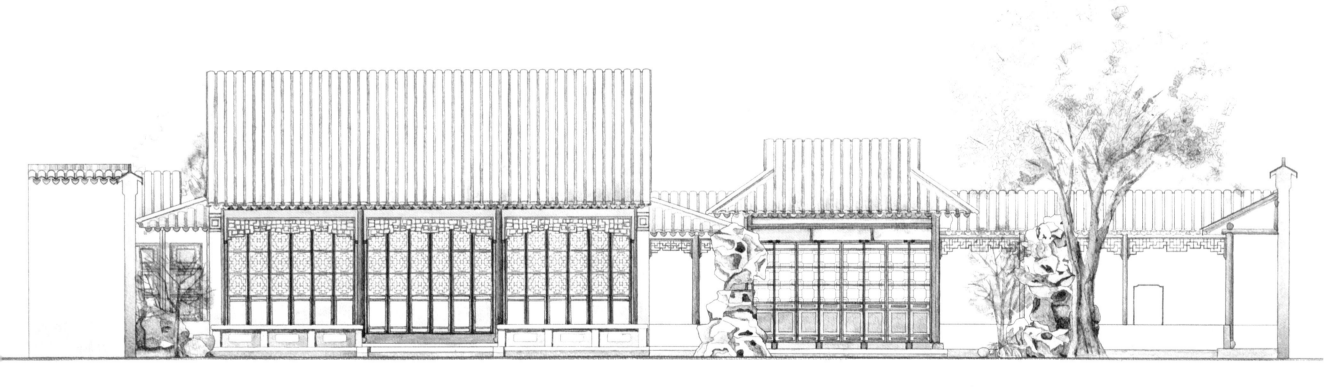

苏州耦园－织帘老屋与鹤寿亭院落南立面图

South elevation: Zhilian Laowu and Heshou Ting, Ou Yuan, Suzhou

苏州沧浪亭 – 沿河北立面图
North elevation: Riverside, Canglang Ting, Suzhou

苏州沧浪亭 – 全园南北剖面图（东视）
North-south section: Canglang Ting, Suzhou

本书所录五处园林位置索引图
A map showing the locations of the five gardens included in the volume in relation to some other cultural heritage sites in Suzhou

苏州艺圃
Yi Pu (Garden of Cultivation), Suzhou

- 地址：苏州市老城区皋桥吴趋坊文衙弄内
- 年代：明—清
- 保护级别：世界文化遗产，全国重点文物保护单位（第六批）

- Address: Wenya Lane, Wuqu Ward, Gusu Dist., Suzhou
- Origin: 16th century
- Status: UNESCO World Cultural Heritage Site (2000), Key Cultural Heritage Site under State Protection (2006)

01 苏州艺圃

艺圃位于苏州老城西北的文衙弄，由袁祖庚创始于明嘉靖年间（1522—1566年），名"醉颖堂"。明万历十八年（1590年），袁祖庚卒，其后园子逐渐衰败。万历末年，文震孟购得醉颖堂，改名为"药圃"。其卒于明崇祯九年（1636年）后，药圃于明末的战乱中有若干建筑被毁，占地面积亦有缩减。清顺治十六年（1659年）末，流寓苏州的山东人姜埰购得药圃为宅，改名"颐圃"，后来其子姜实节改名为"艺圃"，此名沿用至今。今天所见艺圃的基本格局和主要建筑即是在姜氏为园主时奠定的。

姜氏之后，艺圃屡易其主，百余年间规模不断缩小。清道光十九年（1839年），丝绸商人胡寿康、张如松为创建丝绸同业会馆购买了艺圃，命名为"七襄公所"。这一时期艺圃得到了全面的整修并增建了若干建筑，形成了今天的基本面貌。艺圃总占地面积现近六亩，以池（面积近一亩）为主体。主要建筑环池布列：池北有博雅堂和水榭（现名延光阁）；池东有旸谷书堂和爱莲窝，书堂东为世纶堂和东莱草堂，东莱草堂东邻为铚钰斋小院，爱莲窝以南有乳鱼亭伸入池中，正南有思嗜轩；池西有响月廊，其南端通向浴鸥小园及香草居、南斋一组建筑。

艺圃于1963年列为苏州市文物保护单位，1995年提升为江苏省文物保护单位。2000年，艺圃作为"苏州古典园林"的增补项目列入《世界遗产名录》。2006年，列入第六批全国重点文物保护单位。

图1：博雅堂
Fig.1: Boya Tang

图2：池与水榭
Fig.2: Yanguang Ge and the Pond

主要参考文献：
[1] [明] 姜埰. 敬亭集 [M]. 上海：华东师范大学出版社，2011.
[2] [清] 张廷玉等. 明史 [M]. 北京：中华书局，1974.
[3] [清] 汪琬. 尧峰文钞 [M]. 五十卷. 文渊阁四库全书本.
[4] [清] 黄之隽. 乾隆江南通志 [M].（中国地方志集成·江南省志辑.第三至六册）. 据乾隆元年刻本影印. 二百卷首四卷. 南京：凤凰出版社，2011.
[5] 曹允源，李根源. 民国吴县志 [M].（中国地方志集成·江苏府县志辑.第十一至十二册）. 据民国二十二年苏州文新公司铅印本影印. 八十卷. 江苏古籍出版社，1991.
[6] [清] 杨文荪. 七襄公所记 [M]. 程荃. 题.（碑拓）
[7] 林源，冯珊珊. 苏州艺圃营建考. 中国园林 [J]. 2013, 05:116-119.
[8] 林源. 王石谷《艺圃图》、汪琬"艺圃二记"与苏州艺圃 [J]. 建筑师，2013（12）：92-98.
[9] 林源，张文波. 苏州艺圃 [M]. 北京：中国建筑工业出版社，2017.

测　　绘：
西安建筑科技大学建筑学院景观建筑学2008级——
张霄、吴碧晨、吴迪、卫泽民、张勤、王泳文；冯若文、高丽敏、高轶、蒋勤欣、孙佳楠、任达；何政锐、张勇、于广利、郑科、曾黛林、王晓洁；李化贝、李伟、刘玲、刘腾潇、袁舒；夏颖、孙燕杰、邓怀宇、吕安、张斌、郭润泽；
建筑历史与理论研究生2010级——
刘虹、崔兆瑞、冯珊珊、赖祺彬、岳岩敏、张文波、雷繁、唐浩川、徐蕊、朱庭枢、张昱超、王晓静、谷瑞超；

指导教师：林源
图纸绘制、整理：张文波、陈斯亮、王茹悦、马英晨；
测绘时间：2011年10月

01. Yi Pu (Garden of Cultivation), Suzhou

Located at the northwest part of the historic urban area of Suzhou, the garden, originally named Zuiying Tang (Intoxicated-Wisdom Hall), was originally built by Yuan Zugeng (1519–1590) when Emperor Shizong of Ming (r. 1522–1566) was on the throne. After Yuan died in 1590, the garden gradually fell into decline until purchased and renamed to Yao Pu (Herb Garden) by Wen Zhenmeng (1574–1636) in late Emperor Shenzong's reign (1573–1620). Wen's death coincided with the chaotic collapse of Ming Empire (1368–1622), during which the garden saw several structures in it razed and its perimeters withdrawn. In 1659, Jiang Cai (1607–1673), a sojourner from Shandong Province bought the place for his residence and changed the name to Yi Pu[1] (Garden of Recuperation), only for his son Jiang Shijie to rename it again to the current one. The layout of and major structures in the garden we presently see are largely legacies of the Jiangs.

The acreage of the garden continued to shrink in the next one hundred or so years as it passed through the possessions of various owners. In 1893, Hu Shoukang and Zhang Rusong, both silk merchants, bought the place for a silk trading guild with the new name Qixiang Gongsuo (Weaver Girl Guild)[2]. It was during this period that a relatively thorough rehabilitation, plus an addition of several structures, happened to the garden, conferring on it an appearance roughly resembling what we see today. The garden now has an approx. 4000m² layout spatially dominated by the nearly 670m² pond in the center. Boya Tang (Erudition Hall) and Yanguang Ge (Inherited-Honor Pavilion), a waterside pavilion, are located to the north of the pond, while Yanggu Shutang (Sunrise Study) and Ailian Wo (Lotus-Loving Study) to its east. Shilun Tang (Statecraft Hall) and Donglai Caotang (Hermit Thatched-Cottage) lie to the east of the two studies, while a small courtyard featuring Botuo[3] Zhai (Botuo Study) is on the east side of Donglai Caotang. To the south of Ailian Wo, Ruyu Ting (Baby Fish Pavilion) protrudes over the pond and on its due south stands Sishi Xuan (Jujube Pavilion). Along the west bank of the pond runs Xiangyue Lang (Moonlight Enjoying Gallery) which leads south to another group of structures including Xiangcao Ju (Fragrant-Plant Chamber), Nan Zhai (South Study), etc. In 1963, the City of Suzhou listed the Garden as a Cultural Heritage Site under Municipal Protection and in 1995 Jiangsu Province upgraded it to the Provincial Level. In 2000, it was included, as part of an extension, to the Classical Gardens of Suzhou, an inscribed UNESCO World Cultural Heritage Site. In 2006 it joined the ranks of Key Cultural Heritage Sites under State Protection (6th Round of Nomination).

图 3：南侧池山
Fig.3: South bank of the Pond

图 4：浴鸥小院
Fig.4: Yuou Courtyard

图 5：自世伦堂望东莱草堂
Fig.5: A west view of Donglai Caotang from Shilun Tang

Notes:

1. The name Yi Pu (Garden of Recuperation) is homophonic to, yet in different Chinese characters from, the garden's present-day name.

2. Since historical Chinese place-names tended to be excessively allusive and archaic, heavy loss in translation to English would be inevitable. This book, with no intention of fully deciphering these names, in its English text refers to them mainly with transliterations accompanied by bracketed explanations in a paraphrasing manner.

3. Botuo was a historical type of flour-made dim-sum that may have been similar to noodles or dumplings in different areas of China.

Bibliography

[1] JIANG, C. (2001) *Jingting Ji* (Collected Works of Jing-Ting). Shanghai: Normal East University Press (in traditional Chinese).

[2] ZHANG, T. (1974) *Ming Shi* (A History of Ming). Beijing: Zhonghua Book Company (in traditional Chinese).

[3] WANG, W. *Yaofeng Wenchao* (Selected Works of Yao-Feng) (Siku Quanshu Edition, electronic version), consulted on Oct 01, 2016 (in traditional Chinese).

[4] HUANG, Z. (2011) *Jiangnan Tongzhi* (Compiled Records of Jiangnan). Nanjing: Phoenix Press (in traditional Chinese).

[5] CAO, Y. and LI, G. (1991) *Minguo Wuxian Zhi* (ROC Compiled Records for the County of Wu). Nanjing: Jiangsu Guji Press (in traditional Chinese).

[6] YANG, W. *Qixiang Gongsuo Ji* (On Qixiang Guild) (a tablet rubbing), consulted on Oct 01, 2016 (in traditional Chinese).

[7] LIN, Y. & FENG S. (2013) 'Study on the history of Yipu Garden in Suzhou'. in *Chinese Landscape Architecture*, 2013, 5: pp. 116-119 (in simplified Chinese).

[8] LIN, Y. (2013) *The picture of Yipu by Wangshigu, The Notes of Yipu by Wang Wan and Yipu in Suzhou. The Architect*, 2013, 12: pp. 92-98 (in Simplified Chinese).

[9] LIN,Y. and Zhang W. (2017) *Yipu in Suzhou*. Beijing: China Architecture & Building Press, 2017 (in Simplified Chinese).

Credits

Surveyors:

2013 Undergraduate Class of Landscape Architecture (in groups):

ZHANG Xiao, WU Bichen, WU Di, WEI Zemin, ZHANG Qin, WANG Yongwen, FENG Ruowen, GAO Limin, GAO Yi, JIANG Qinxin, SUN Jianan, REN Da; HE Zhengrui, ZHANG Yong, YU Guangli, ZHENG Ke, ZENG Dailin, WANG Xiaojie; LI Wei, LI Huabei, LIU Ling, LIU Tengxiao, YUAN Shu; XIA Ying, SUN Yanjie, DENG Huaiyu, LV An, ZHANG Bin, GUO Runze

2013 Postgraduate Class of Architectural History:

LIU Hong, CUI Zhaorui, FENG Shanshan, LAI Qibin, YUE Yanmin, ZHANG Wenbo, LEI Fan, TANG Haochuan, XU Rui, ZHU Tingshu, ZHANG Yuchao, WANG Xiaojing, GU Ruichao

Tutor: LIN Yuan

Drawing Preparation: ZHANG Wenbo, CHEN Siliang, WANG Ruyue, MA Yingchen

Date: October, 2011

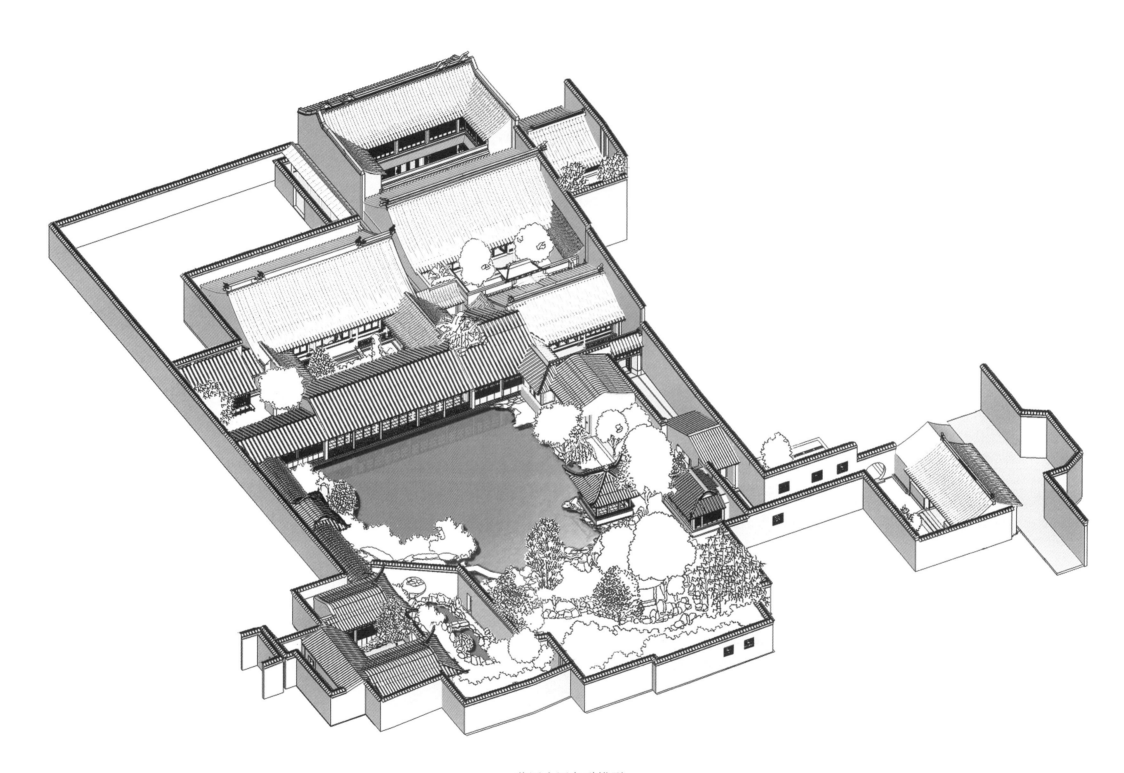

艺圃全园鸟瞰模型
A bird-view model: Yi Pu

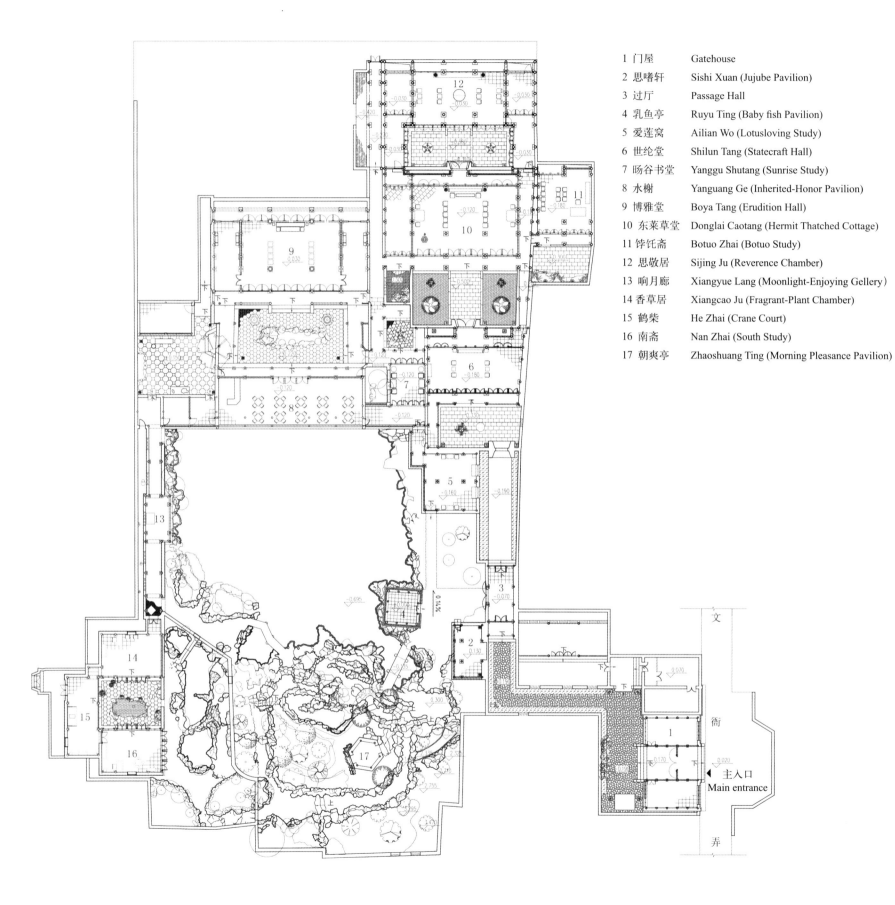

植物配置表
Vegetation

植物图例 Legend	植物名称 Binomial	植物科属 Family
	白皮松 Pinus bungeana	松科松属
	木犀 Osmanthus fragrans	木犀科木犀属
	银杏 Ginkgo biloba	银杏科银杏属
	朴树 Celtis sinensis	榆科朴属
	香橼 Citrus medica	芸香科柑橘属
	枳椇 Hovenia acerba	鼠李科枳椇属
	榆树 Ulmus pumila	榆科榆属
	鹅掌楸 Liriodendron chinense	木兰科鹅掌楸属
	鸡爪槭 Acer palmatum	槭树科槭属
	蜡梅 Chimonanthus praecox	腊梅科腊梅属
	紫荆 Cercis chinensis	豆科紫荆属
	黄杨 Buxus sinica	黄杨科黄杨属

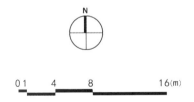

总平面图
Site plan: Ground level

1	门屋	Gatehouse
2	思嗜轩	Sishi Xuan (Jujube Pavilion)
3	过厅	Passage Hall
4	乳鱼亭	Ruyu Ting (Baby fish Pavilion)
5	爱莲窝	Ailian Wo (Lotusloving Study)
6	世纶堂	Shilun Tang (Statecraft Hall)
7	旸谷书堂	Yanggu Shutang (Sunrise Study)
8	衍榭	Yanguang Ge (Inherited-Honor Pavilion)
9	博雅堂	Boya Tang (Erudition Hall)
10	东莱草堂	Donglai Caotang (Hermit Thatched Cottage)
11	铇钰斋	Botuo Zhai (Botuo Study)
12	思敬居	Sijing Ju (Reverence Chamber)
13	响月廊	Xiangyue Lang (Moonlight-Enjoying Gellery)
14	香草居	Xiangcao Ju (Fragrant-Plant Chamber)
15	鹤柴	He Zhai (Crane Court)
16	南斋	Nan Zhai (South Study)
17	朝爽亭	Zhaoshuang Ting (Morning Pleasance Pavilion)

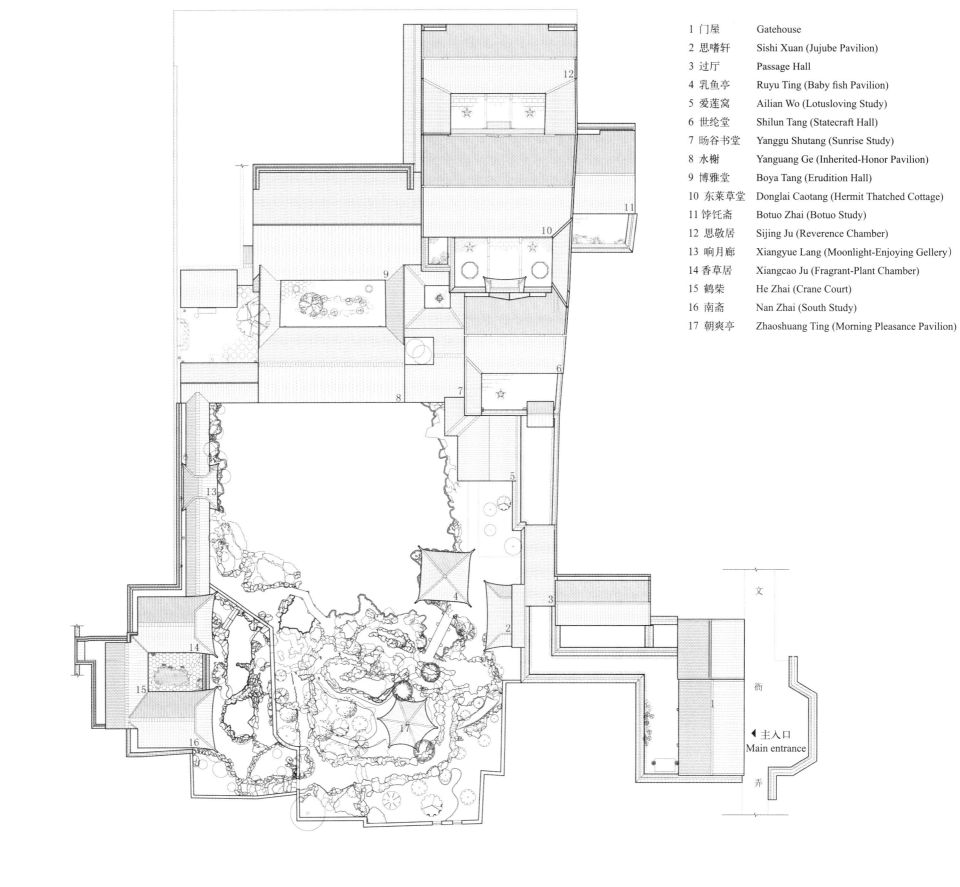

1	门屋	Gatehouse
2	思嗜轩	Sishi Xuan (Jujube Pavilion)
3	过厅	Passage Hall
4	乳鱼亭	Ruyu Ting (Baby fish Pavilion)
5	爱莲窝	Ailian Wo (Lotusloving Study)
6	世纶堂	Shilun Tang (Statecraft Hall)
7	旸谷书堂	Yanggu Shutang (Sunrise Study)
8	水榭	Yanguang Ge (Inherited-Honor Pavilion)
9	博雅堂	Boya Tang (Erudition Hall)
10	东莱草堂	Donglai Caotang (Hermit Thatched Cottage)
11	铎铊斋	Botuo Zhai (Botuo Study)
12	思敬居	Sijing Ju (Reverence Chamber)
13	响月廊	Xiangyue Lang (Moonlight-Enjoying Gellery)
14	香草居	Xiangcao Ju (Fragrant-Plant Chamber)
15	鹤柴	He Zhai (Crane Court)
16	南斋	Nan Zhai (South Study)
17	朝爽亭	Zhaoshuang Ting (Morning Pleasance Pavilion)

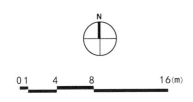

0 1　4　8　16(m)

屋顶平面图
Site plan: Roof level

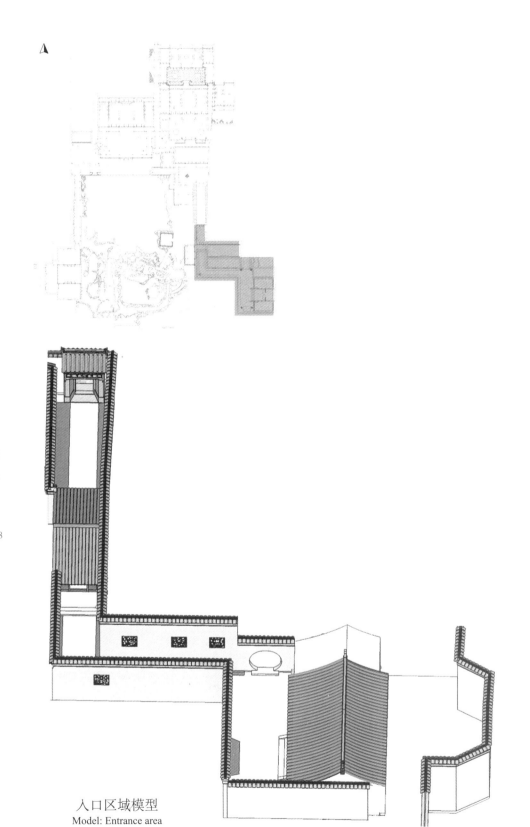

入口区域模型
Model: Entrance area

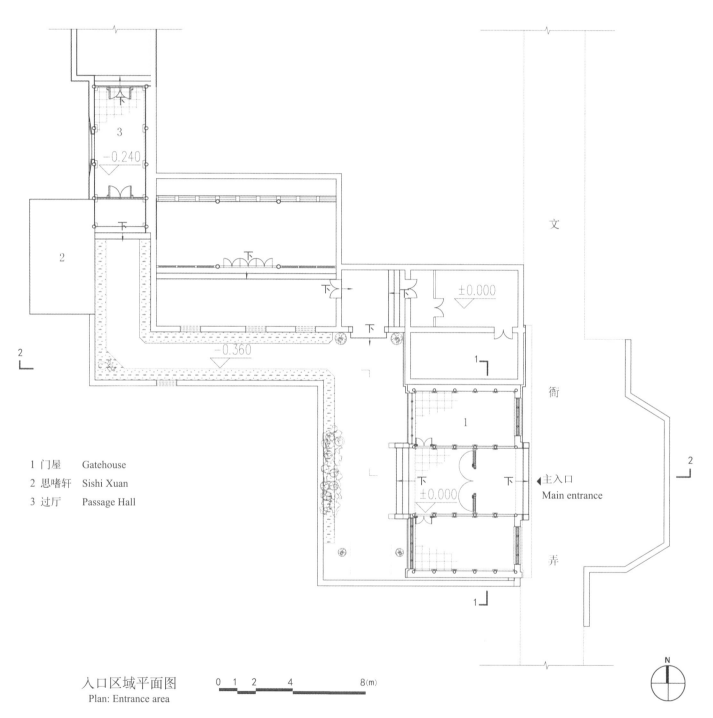

1 门屋　Gatehouse
2 思嗜轩　Sishi Xuan
3 过厅　Passage Hall

入口区域平面图
Plan: Entrance area

门屋东立面图

East elevation: Gatehouse

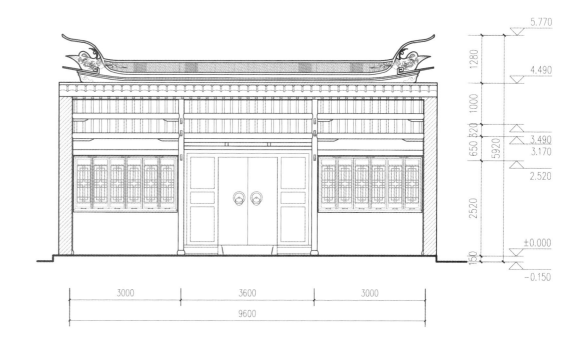

门屋 1-1 剖面图

Section 1-1: Gatehouse

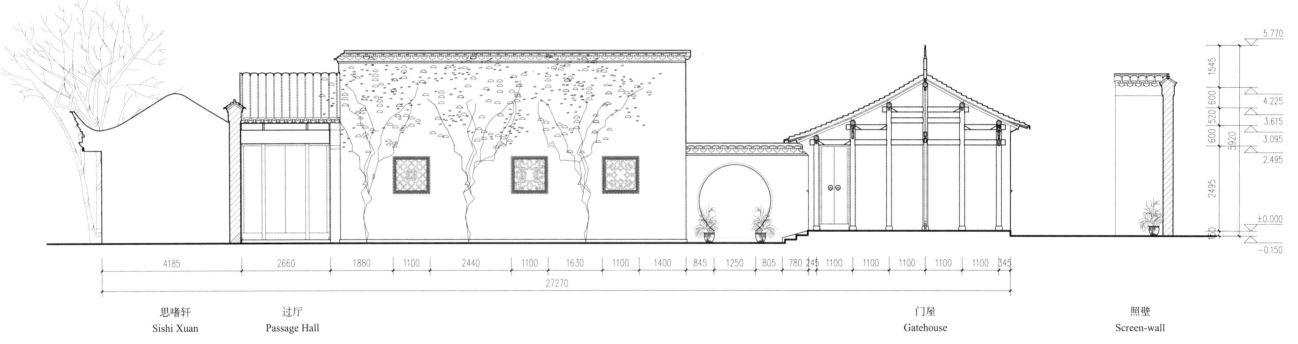

入口区域 2-2 剖面图
Section 2-2: Entrance area

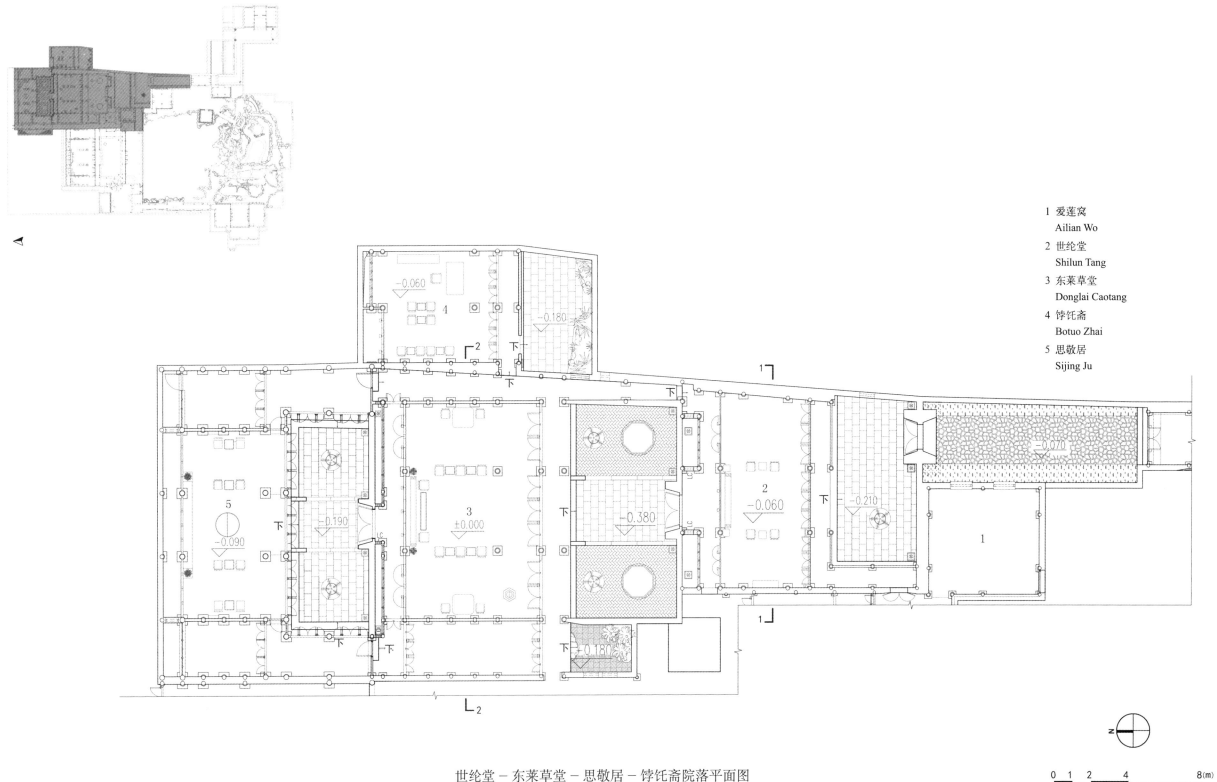

1	爱莲窝 Ailian Wo
2	世纶堂 Shilun Tang
3	东莱草堂 Donglai Caotang
4	饽饦斋 Botuo Zhai
5	思敬居 Sijing Ju

世纶堂－东莱草堂－思敬居－饽饦斋院落平面图
Plan: Shilun Tang, Donglai Caotang, Sijing Ju and Botuo Zhai

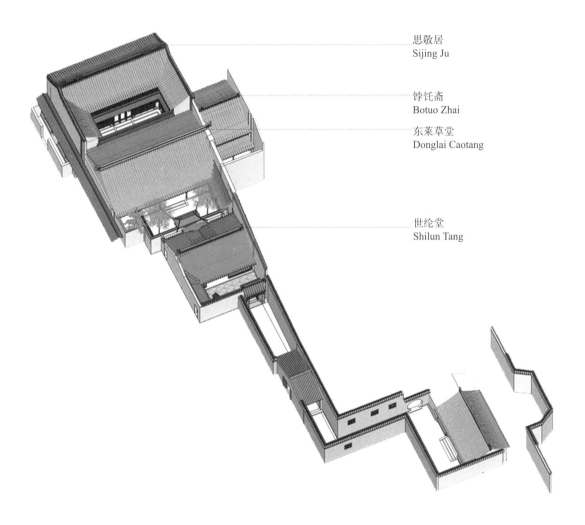
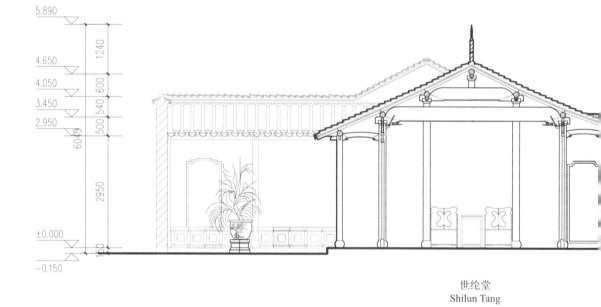

思敬居 Sijing Ju
饽饦斋 Botuo Zhai
东莱草堂 Donglai Caotang
世纶堂 Shilun Tang

世纶堂 Shilun Tang

世纶堂－东莱草堂－思敬居－饽饦斋院落模型
Model: Shilun Tang, Donglai Caotang, Sijing Ju and Botuo Zhai

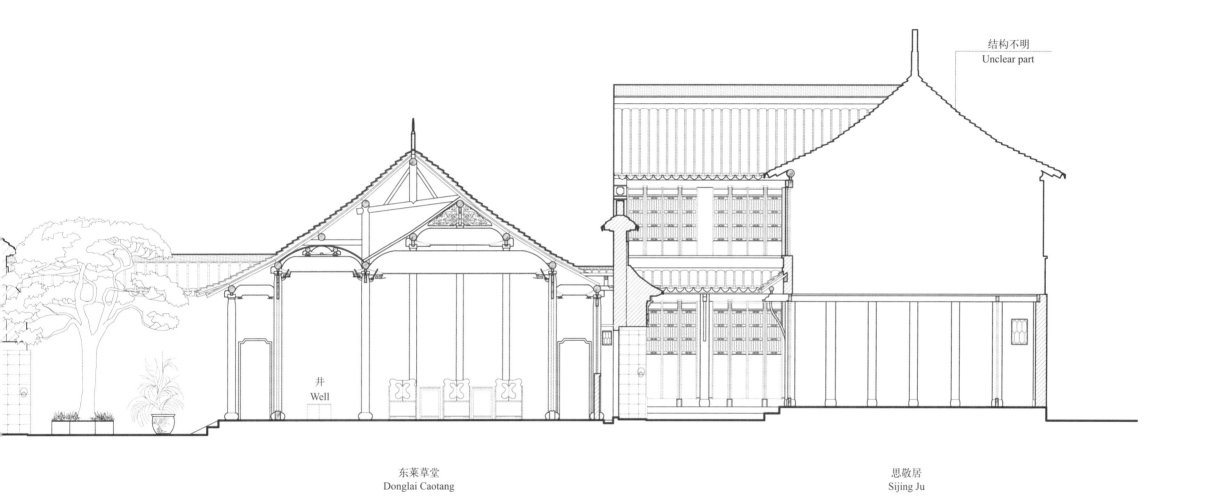

世纶堂－东莱草堂院落南北剖面图（西视）
South-north section: Shilun Tang and Donglai Caotang

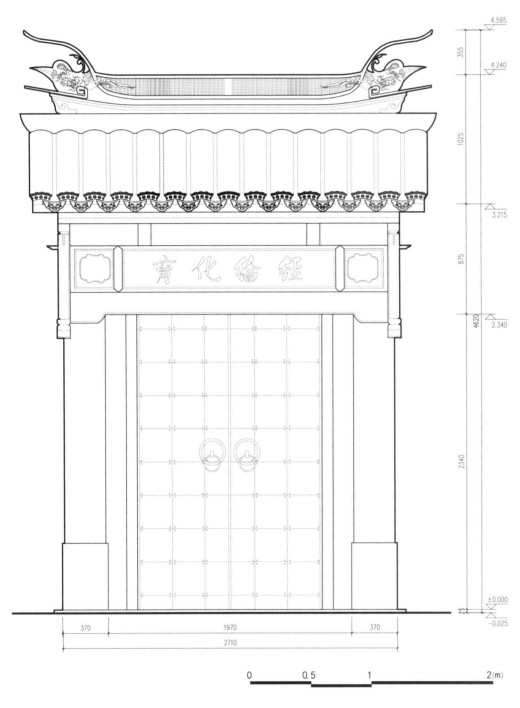

"经纶化育"墙门南立面图
South elevation: Gate bearing the inscription 'Jinglun Huayu' (Statecraft and Cultivation)

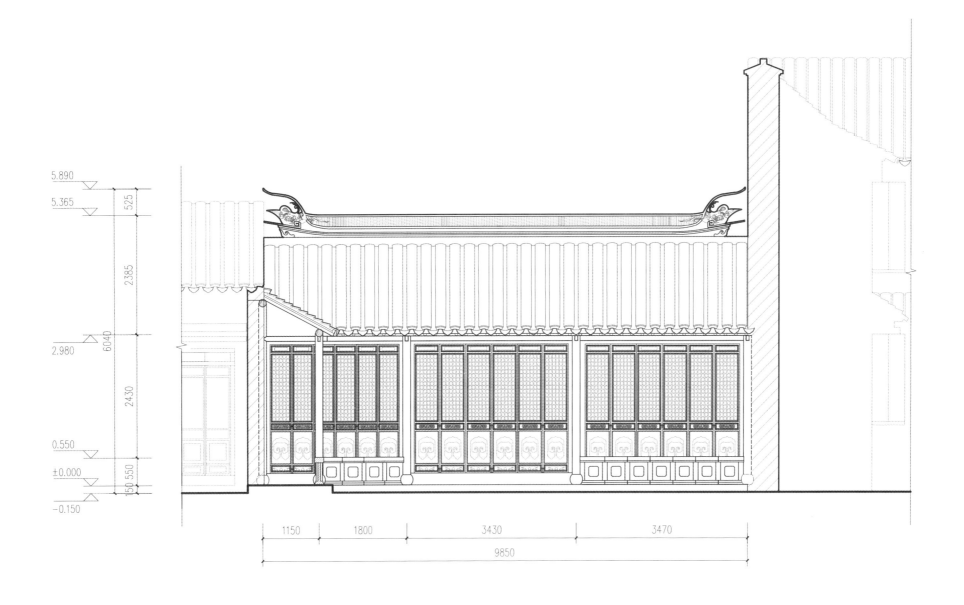

世纶堂南立面图
South elevation: Shilun Tang

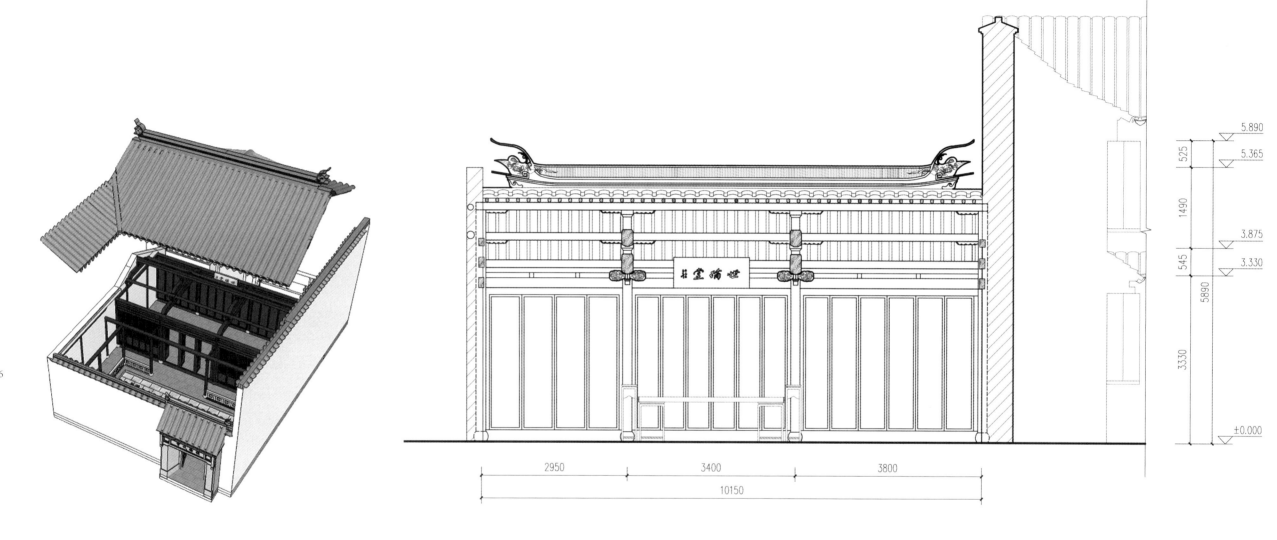

世纶堂院落模型
Model: Shilun Tang

世纶堂 1-1 剖面图
Section 1-1: Shilun Tang

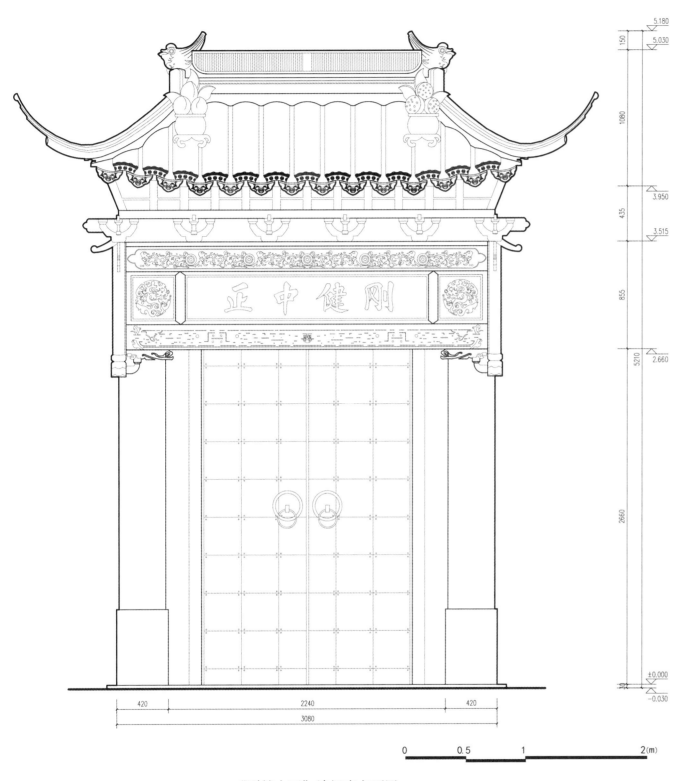

"刚健中正"墙门南立面图

South elevation: The gate bearing the inscription 'Gangjian Zhongzheng' (Resilience and Righteousness)

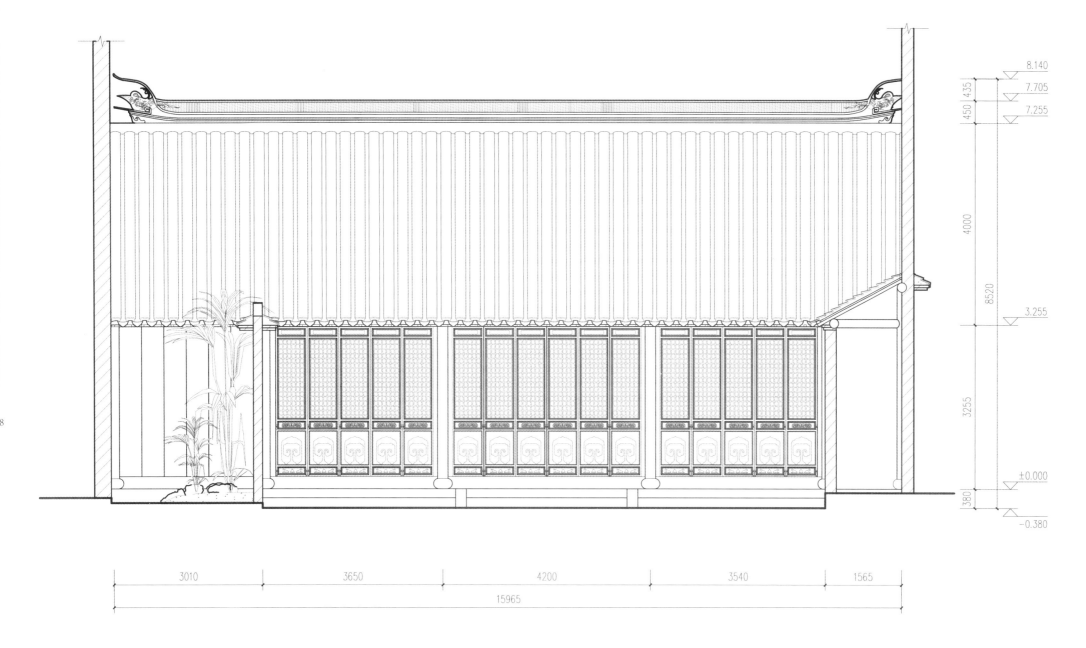
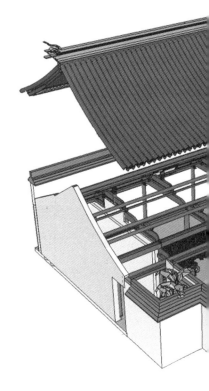

东莱草堂南立面图
South elevation: Donglai Caotang

东莱草堂院落模型
Model: Donglai Caotang

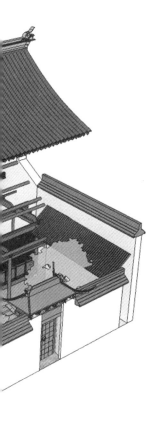
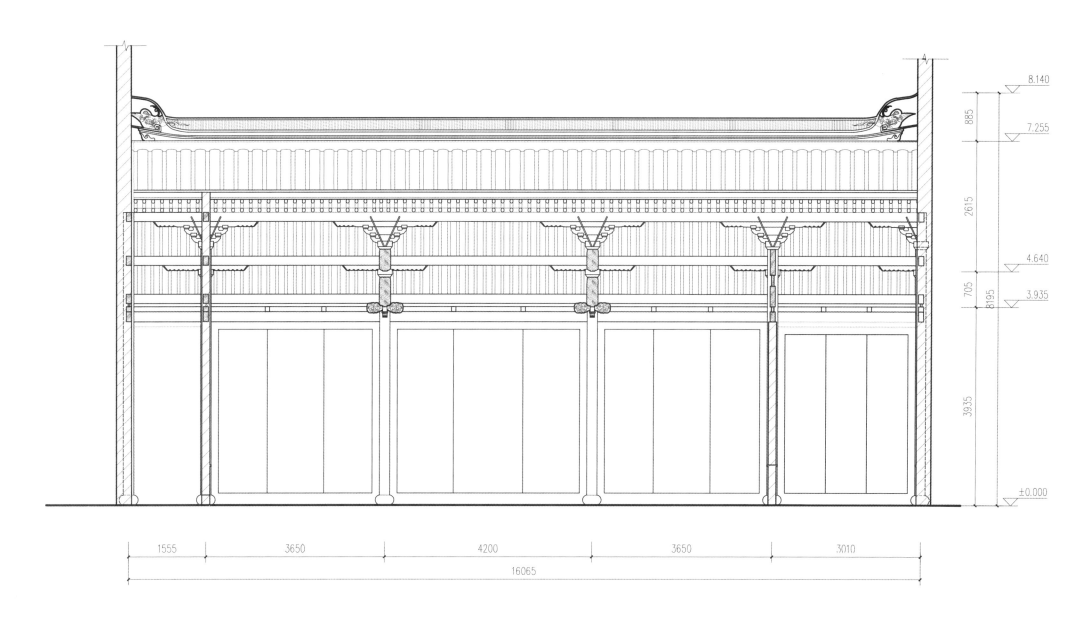

东莱草堂 2-2 剖面图
Section 2-2: Donglai Caotang

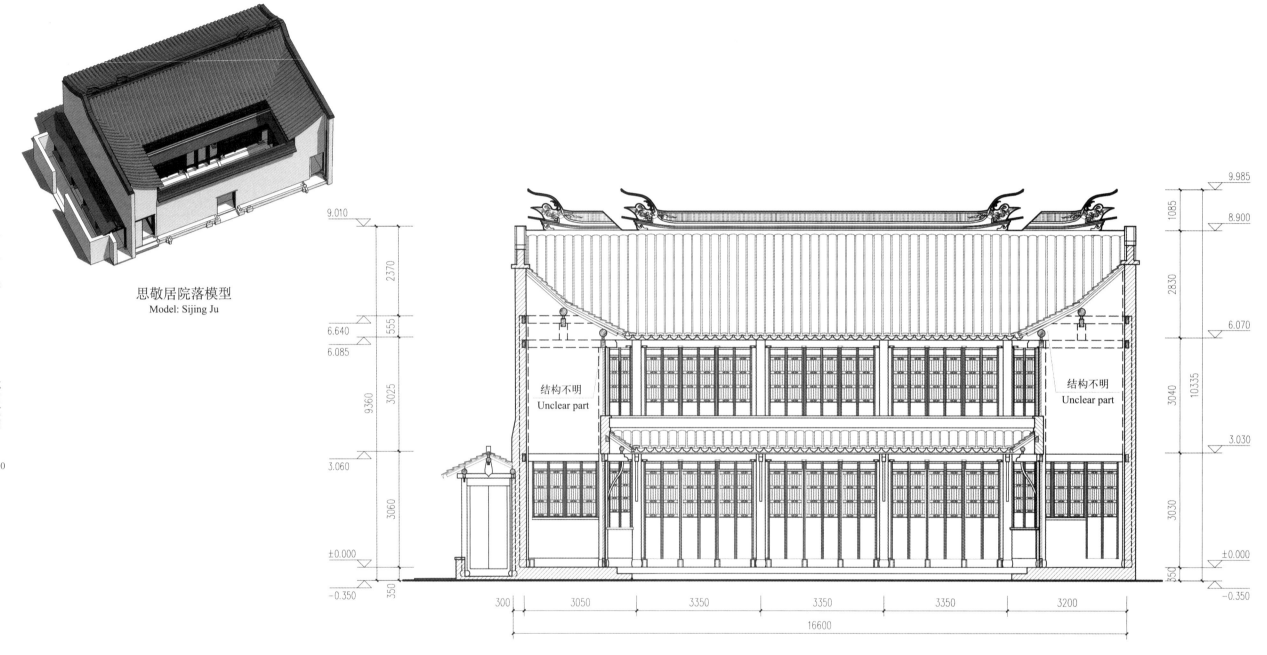

思敬居院落模型
Model: Sijing Ju

思敬居南立面图
South elevation: Sijing Ju

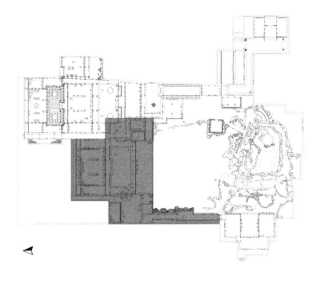
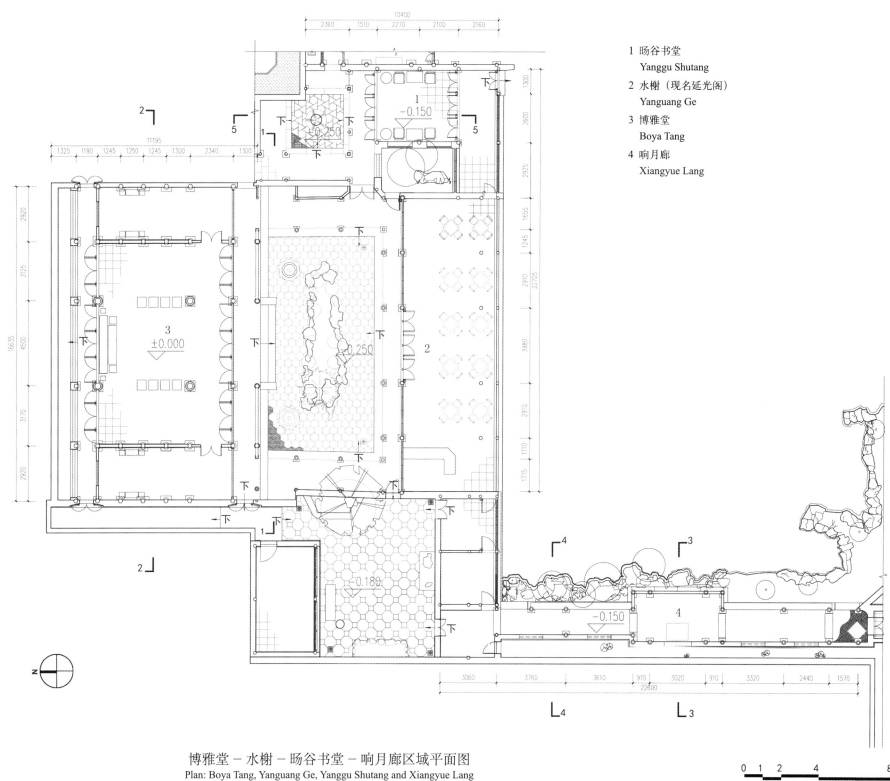

1 旸谷书堂　Yanggu Shutang
2 水榭（现名延光阁）　Yanguang Ge
3 博雅堂　Boya Tang
4 响月廊　Xiangyue Lang

博雅堂－水榭－旸谷书堂－响月廊区域平面图
Plan: Boya Tang, Yanguang Ge, Yanggu Shutang and Xiangyue Lang

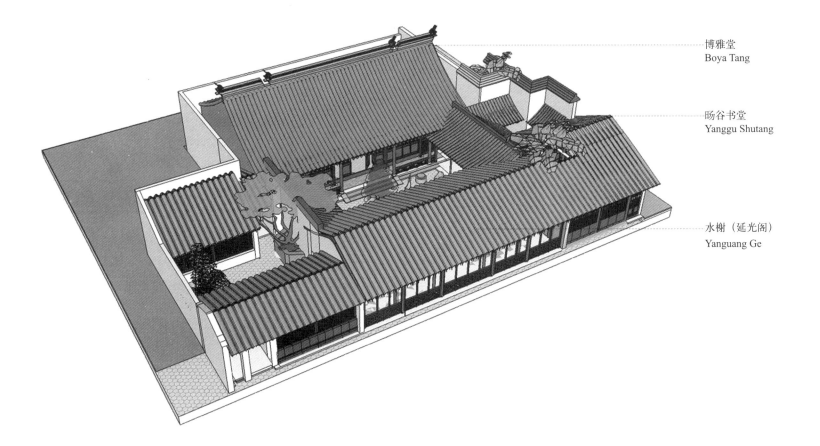

博雅堂 — 水榭 — 旸谷书堂区域模型
Model: Boya Tang, Yanguang Ge and Yanggu Shutang

博雅堂 Boya Tang
旸谷书堂 Yanggu Shutang
水榭（延光阁）Yanguang Ge

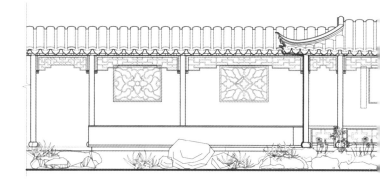

响月廊 Xiangyue Lang

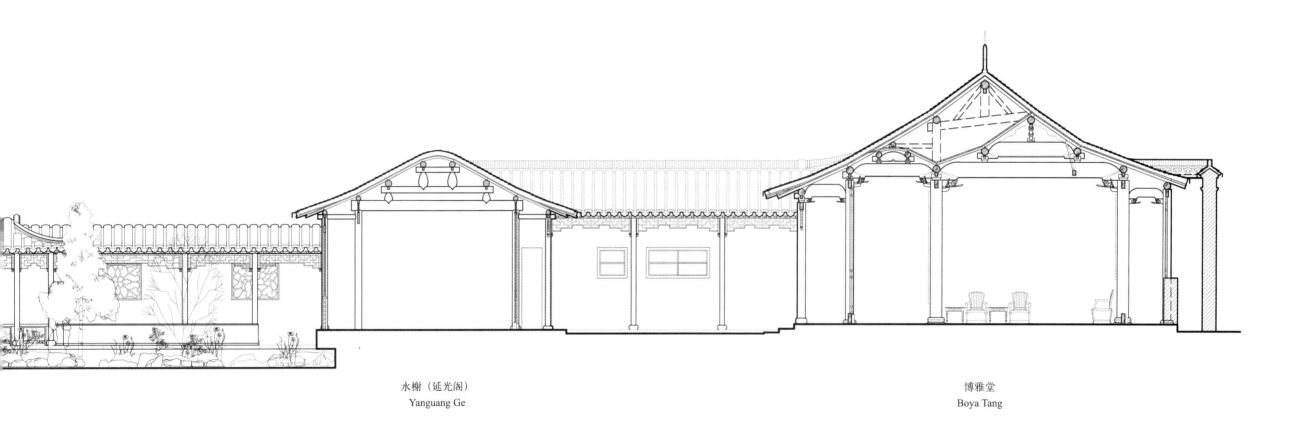

水榭（延光阁）
Yanguang Ge

博雅堂
Boya Tang

博雅堂－水榭（现名延光阁）－旸谷书堂－响月廊区域南北剖面图（西视）
South-north section: Boya Tang, Yanguang Ge, Yanggu Shutang and Xiangyue Lang

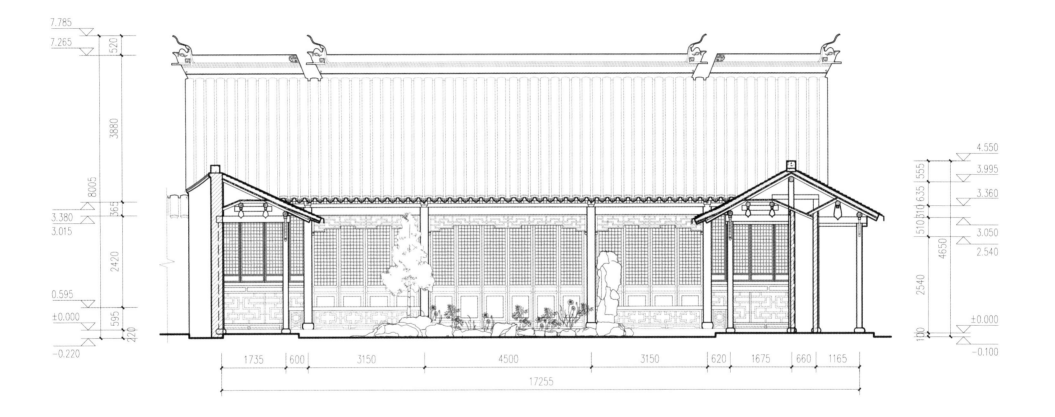

博雅堂院落 1-1 剖面图
Section 1-1: Boya Tang

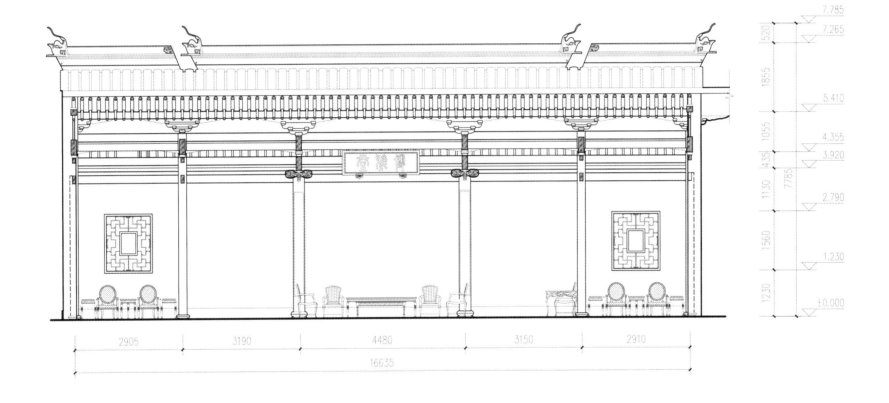

博雅堂院落 2-2 剖面图
Section 2-2: Boya Tang

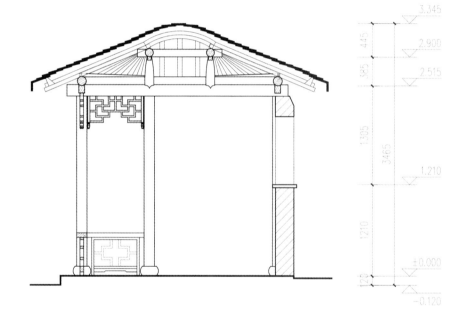
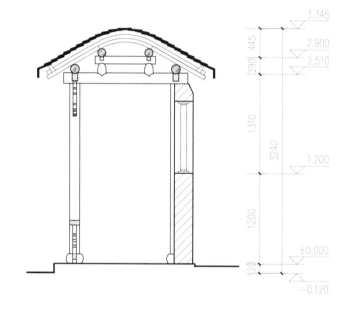

响月廊 3-3 剖面图　　　0　0.5　1　　2(m)
Section 3-3: Xiangyue Lang

响月廊 4-4 剖面图　　　0　0.5　1　　2(m)
Section 4-4: Xiangyue Lang

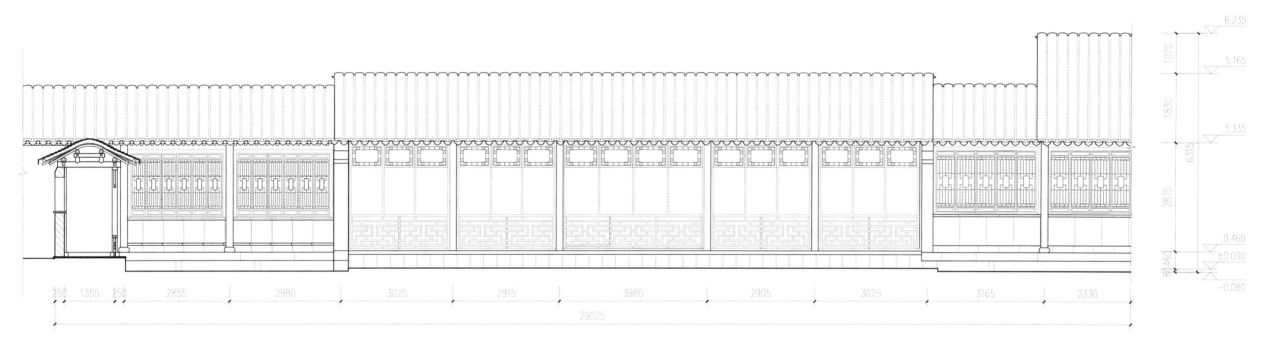

水榭（延光阁）南立面图
South elevation: Yanguang Ge

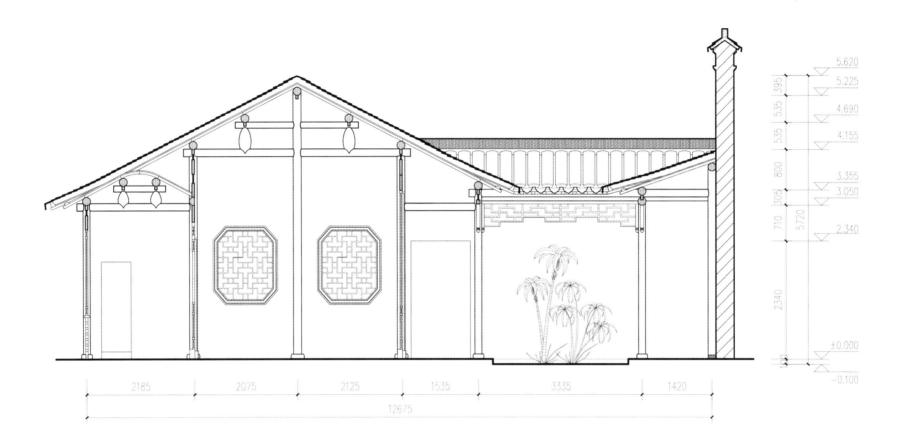

旸谷书堂 5-5 剖面图
Section 5-5: Yanggu Shutang

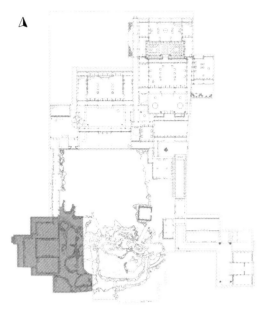
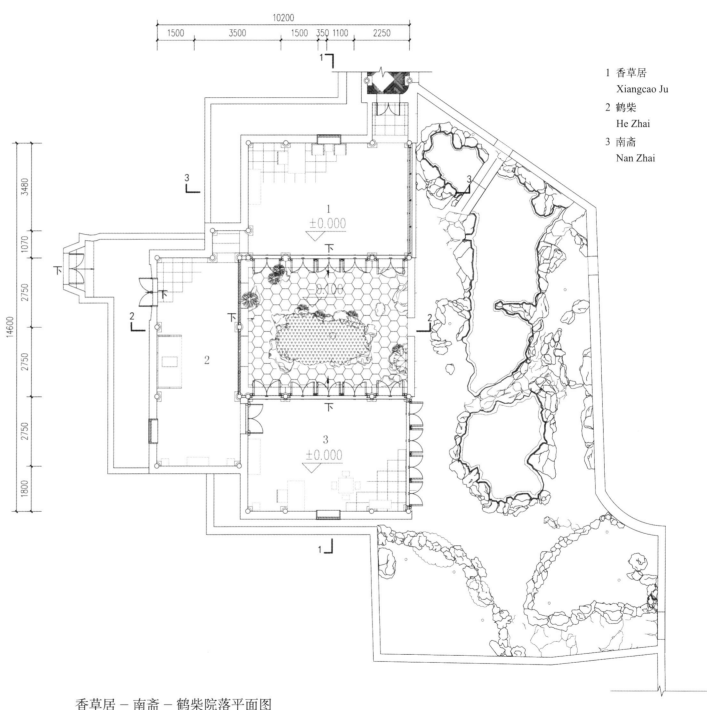
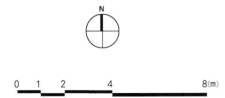

1 香草居 Xiangcao Ju
2 鹤柴 He Zhai
3 南斋 Nan Zhai

香草居－南斋－鹤柴院落平面图
Plan: Xiangcao Ju, Nan Zhai and He Zhai

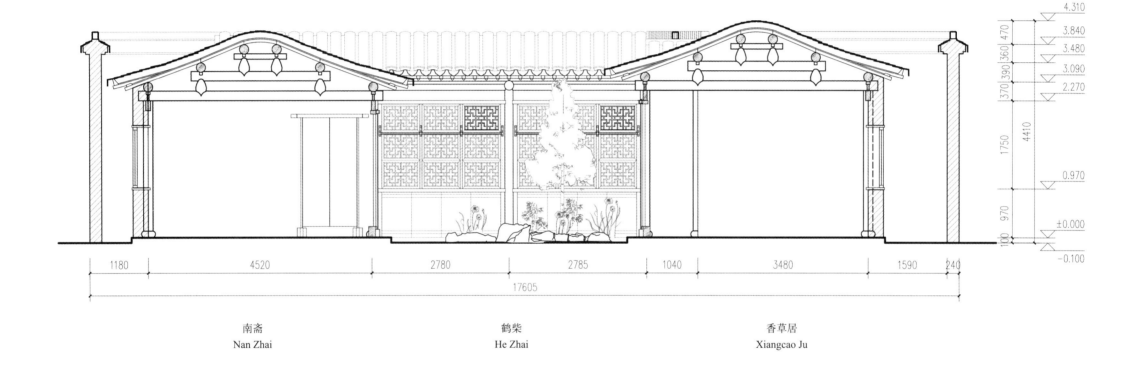

香草居－南斋－鹤柴院落 1-1 剖面图
Section 1-1: Xiangcao Ju, Nan Zhai and He Zhai

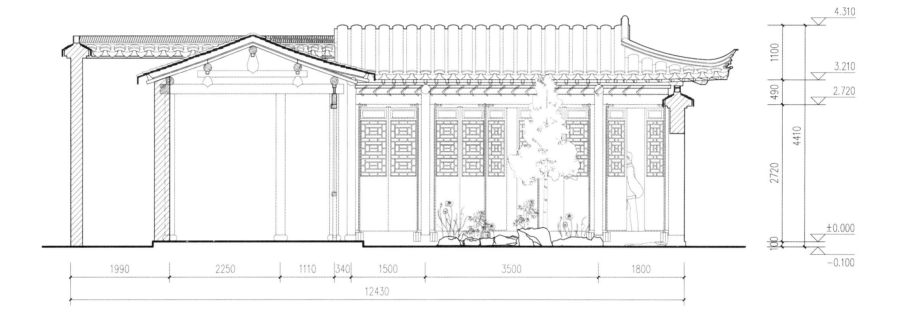

香草居－南斋－鹤柴院落 2-2 剖面图
Section 2-2: Xiangcao Ju, Nan Zhai and He Zhai

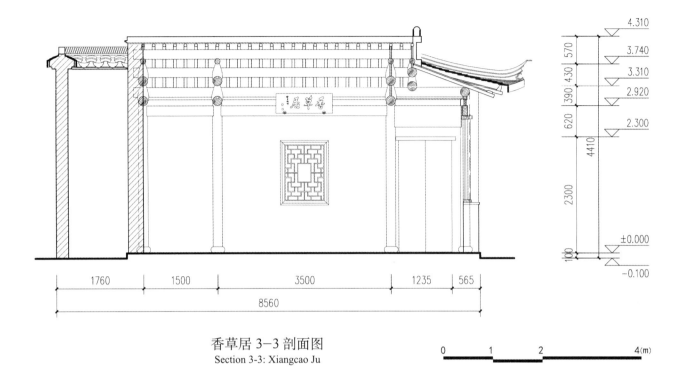

香草居 3-3 剖面图
Section 3-3: Xiangcao Ju

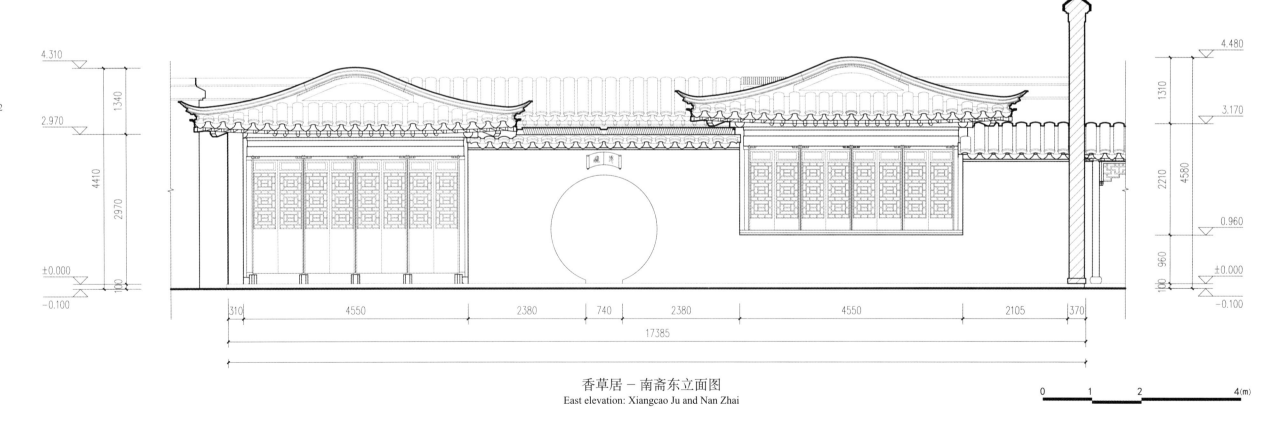

香草居 — 南斋东立面图
East elevation: Xiangcao Ju and Nan Zhai

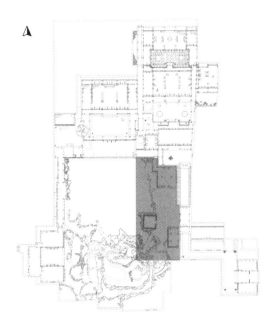

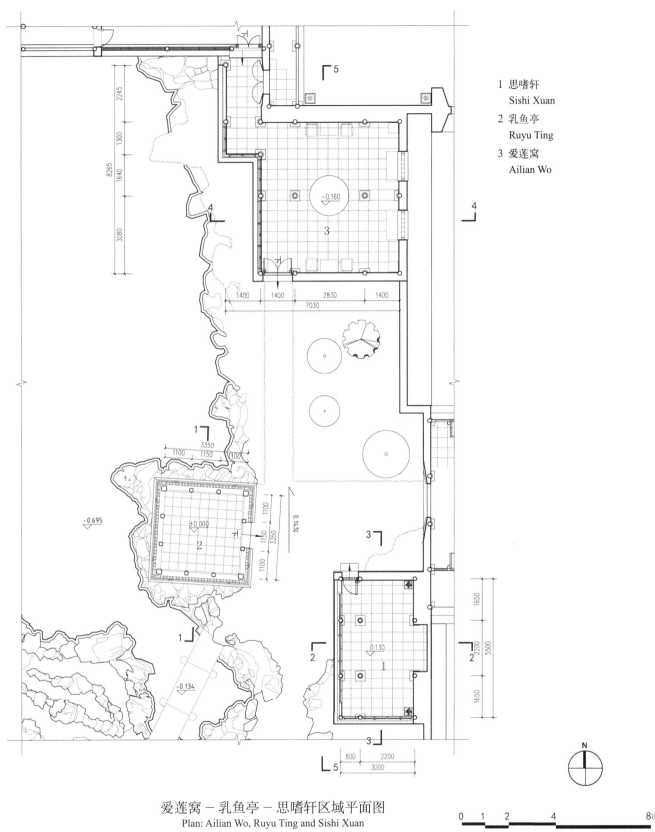

1	思嗜轩 Sishi Xuan
2	乳鱼亭 Ruyu Ting
3	爱莲窝 Ailian Wo

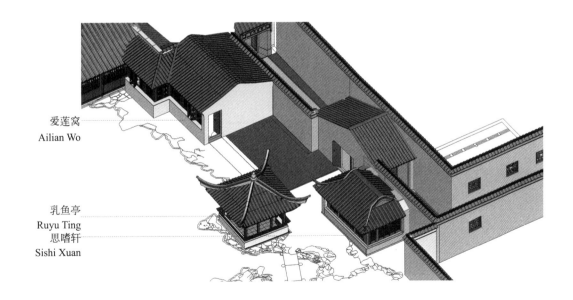

爱莲窝－乳鱼亭－思嗜轩区域模型
Model: Ailian Wo, Ruyu Ting and Sishi Xuan

爱莲窝－乳鱼亭－思嗜轩区域平面图
Plan: Ailian Wo, Ruyu Ting and Sishi Xuan

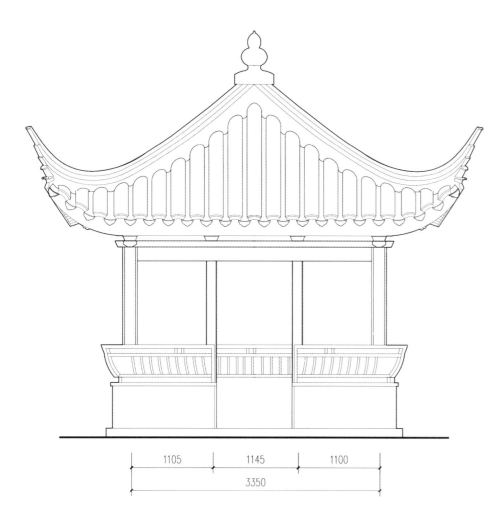

乳鱼亭东立面图
East elevation: Ruyu Ting

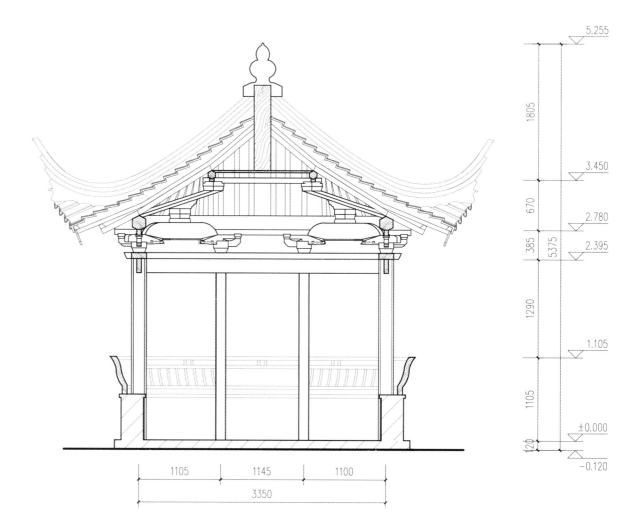

乳鱼亭 1-1 剖面图
Section 1-1: Ruyu Ting

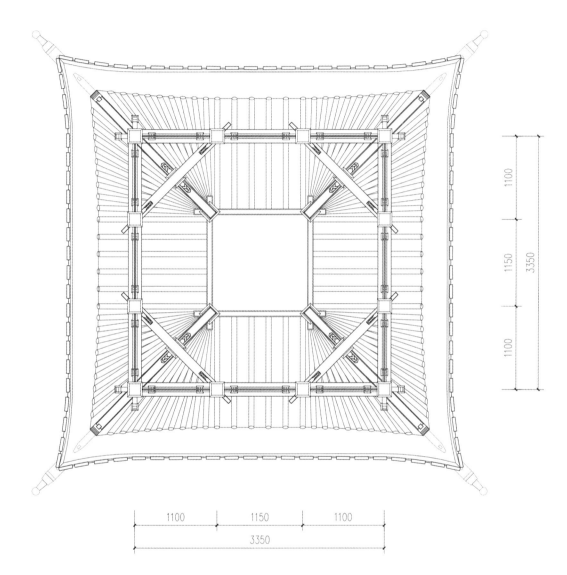

乳鱼亭梁架仰视图
Reflected truss plan: Ruyu Ting

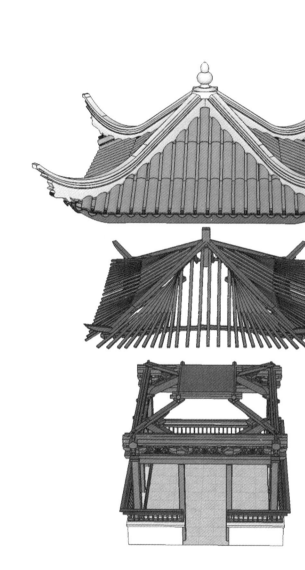

乳鱼亭模型
Model: Ruyu Ting

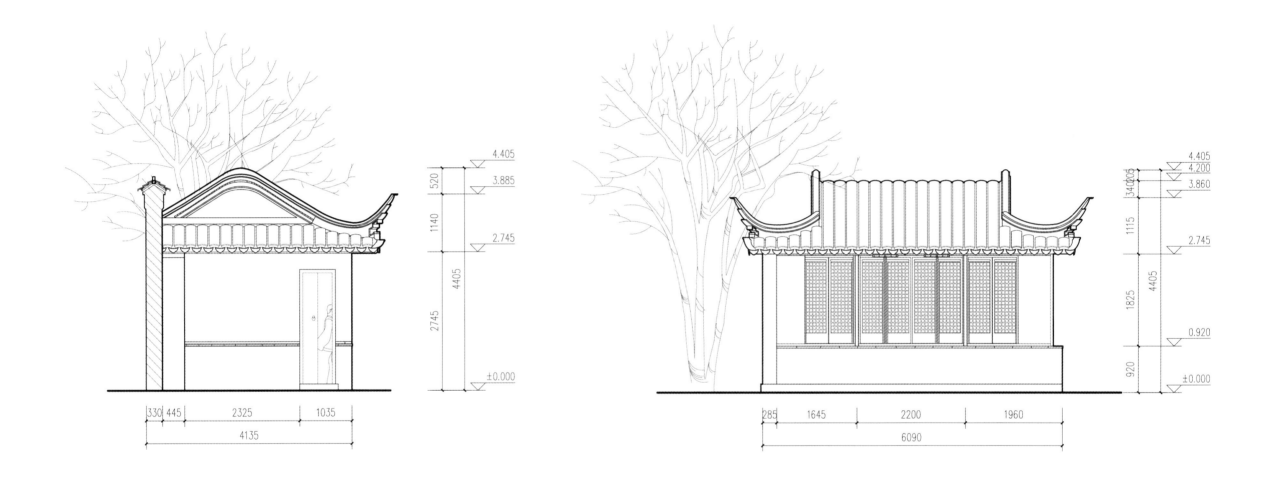

思嗜轩北立面图
North elevation: Sishi Xuan

思嗜轩西立面图
West elevation: Sishi Xuan

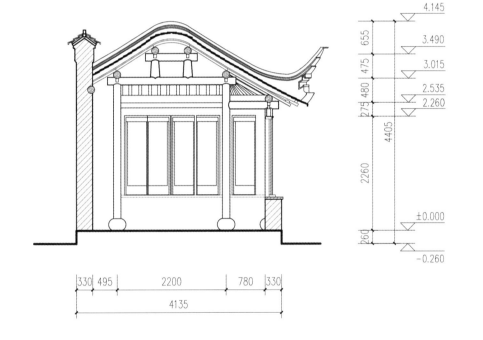

思嗜轩 2-2 剖面图
Section 2-2: Sishi Xuan

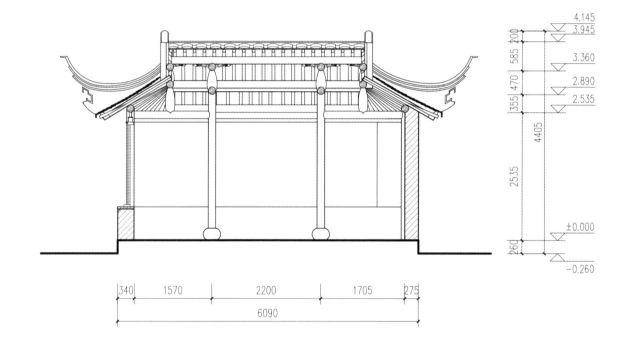

思嗜轩 3-3 剖面图
Section 3-3: Sishi Xuan

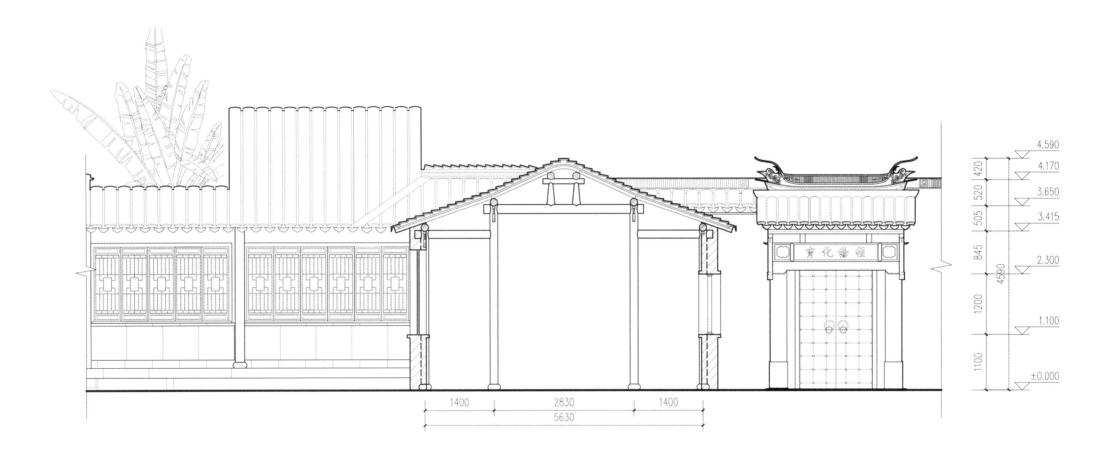

爱莲窝 4-4 剖面图
Section 4-4: Ailian Wo

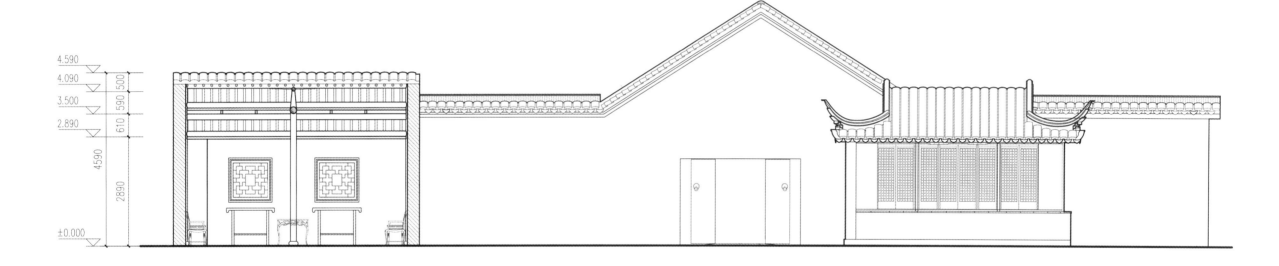

爱莲窝思嗜轩西立面图 5-5 剖面图

Section 5-5: Ailian Wo / West elevation: Sishi Xuan

苏州环秀山庄

Huanxiu Shanzhuang (Mountain Villa with Embracing Beauty), Suzhou

- 地址：苏州市老城区景德路 262 号
- 年代：清
- 保护级别：世界文化遗产，全国重点文物保护单位（第三批）

· Address: 262 Jingde Rd., Gusu Dist., Suzhou
· Origin: 18th Century
· Status: UNESCO World Cultural Heritage Site (1997), Key Cultural Heritage Site under State Protection (1988)

02 苏州环秀山庄

环秀山庄位于苏州老城西北、景德路上、今苏州刺绣博物馆内，南临王鏊祠，西与艺圃相距不远。全园面积约3亩，以假山为主体。此山是清造园大师戈裕良的作品。园子所在原为五代吴越钱氏"金谷园"旧址。宋时为朱长文乐圃，其后屡有兴废。清乾隆年间（1736—1795年）为刑部员外郎蒋楫宅，其时叠石堆山，掘地时得清泉一眼，命名为"飞雪"。泉水汇聚为池。后相继为尚书毕沅、大学士孙士毅宅。孙士毅之孙孙均于嘉庆年间（1796—1820年）请戈裕良叠筑了园内假山。道光二十九年（1847年），汪为仁购得此地，建汪氏宗祠"耕荫义庄"，重修花园更名为"环秀山庄"，又称"颐园"。后经咸丰（1831—1861年）、同治（1862—1874年）年间战事，园毁损严重。光绪年间（1875—1908年）重修。至1949年时，园子仅存有一山、一池及一座"补秋舫"。1984—1985年进行的大规模整修恢复了有穀堂、四面厅等主要建筑，整修也包括加固假山、清理池水、种植树木等内容。山庄于1988年列为第三批全国重点文物保护单位，1997年作为"苏州古典园林"之一列入《世界遗产名录》。

环秀山庄入口东向，入园后北折进入门厅，经东、西廊向北即达有穀堂。过有穀堂向北为四面厅，厅内悬挂有"环秀山庄"匾额。厅北为全园的主山与池。园的西北角为次山，名为"飞雪"。"飞雪"南有问泉亭，池水萦绕于主山与次山之间。问泉亭北、"飞雪"东为补秋山房。山房的东南有亭，名为"半潭秋水一房山"。

主要参考文献：

[1] [清] 冯桂芬. 显志堂稿 [M]. (近代中国史料丛刊续编·第七十九辑). 据光绪二年校邠庐刊本影印. 十二卷. 台北：文海出版社，1974.

[2] [清] 李铭皖, 冯桂芬. 同治苏州府志 [M]. (中国地方志集成·江苏府县志辑·第七至十册). 据光绪九年刻本影印. 一百五十卷首三卷. 南京：江苏古籍出版社，1991.

[3] 刘敦桢. 苏州古典园林 [M], 北京：中国建筑工业出版社，2005.

[4] [清] 金天羽. 颐园记 [A]. 王稼句. 苏州园林历代文抄 [M] 上海：上海三联书店，2008.

测　　绘：西安建筑科技大学建筑学院景观建筑学2009级—
郑旭静、杨旭、孙冰玉、雷凯、李若瑜、高一凡、梁歌、贾文静、李国庆、钟慧敏、沈尔迪、张闻芯、张克华、胡嘉、郭小楠、李苗苗、马倩、马毓、王薇、董骐玮、许婷婷、薛源、徐诗文、杨菲、杨毓婧、丁婉靖、罗维祯、张雯婕

指导教师：林源、岳岩敏；
图纸绘制、整理：汶武娟、王艺博、杨烜子、魏顾；
测绘时间：2012年10月

图1：入口部分
Fig.1: Entrance area

图2：有穀堂
Fig.2: Yougu Tang

02. Huanxiu Shanzhuang (Mountain Villa with Embracing Beauty), Suzhou

The garden lies inside the present-day Suzhou Museum for Embroidery, Jingde Rd. at the northwest part of the historic urban area of Suzhou. It has the Shrine of Wang Ao (1450–1524) in its proximity to the south and is not far from Yi Pu to the west. The nearly 2000m² site is dominated by rockworks created by the virtuoso landscape garden designer Ge Yuliang (1764–1830). In the long course of history the garden crossed paths with various dignitaries. Its origin could date back to the famous 10th-century Jingu Yuan (Garden of Golden Valley), owned by a royal family member of Wuyue Kingdom (907–978) and the 11th-century Le Pu (Garden of Joy) inhabited by an early Song Dynasty (960–1279) Confucian scholar ZHU Changwen (1039–1098), followed by centuries of vicissitudes. Later in 18th century, families of three Qing Dynasty (1636–1912) high officials, namely JIANG Ji, BI Yuan (1730–1797) and SUN Shiyi (1720–1796), resided here successively during Emperor Gaozong's reign (1736–1795) and made significant contribution to the site's present-day appearance. The pond we see today was formed by a spring, named Feixue (Flying Snowflakes), accidentally dug in Jiang's landscaping project and it was SUN Jun, the grandson of SUN Shiyi, that commissioned the work of Ge Yuliang. In 1847, WANG Weiren, after purchasing the plot, built a garden to the northeast of the ancestral shrine for his family and christened it as what we know today while the garden was then also known as Yi Yuan (Garden of Recuperation). The site sustained severe damages owing to the warfare between 1850s and 1870s but received some repair work when Emperor Dezong (1875–1908) was on the throne. By 1949 the garden had been spared with nothing but one rockery hill, one pond and one houseboat named Buqiu Fang (Autumn-Reinforcing Boat). A large scale restoration project, funded by Suzhou Municipal Bureau of Garden and the Research Institute of Embroidery, was launched in 1984 and finished the next year. The project had structures such as Yougu Tang (Good Harvest Hall) and Simian Ting (Quadrilateral Opening Pavilion) restored, rockworks consolidated, pond cleaned up and trees planted. In 1988, the site was listed as a Key Cultural Heritage Site under State Protection (3rd Round of Nomination) and inscribed as one of the Classical Gardens of Suzhou, a UNESCO World Heritage Site in 1997.

The garden opens to the east. Upon entering, one could turn north and walk into the vestibule. The vestibule is flanked by galleries at both sides, which all lead north to Yougu Tang. Heading further north, one will reach Simian Ting, inside which the name plaque of the garden is hanged. Continuing north, one will finally be welcomed by the primary rockery

图3：有榖堂与四面厅之间的小院
Fig.3: Atrium between Yougu Tang and Simian Ting

图4：主山与石板桥
Fig.4: Primary rockery hill and Stone-Slab Bridge

hill of the garden and the pond. There is, at the northwest corner of the garden, also a secondary rockery hill, named Feixue (Flying Snowflakes), of which Wenquan Ting (Spring-Visiting Pavilion) stands to the south. The two hills are connected by the pond between them which has a sinuous layout. To the north of the primary hill and adjoining Feixue on the east stands Buqiu Shanfang (Autumn-Reinforcing Mountain-House), to the southeast of which there is another pavilion named Bantan Qiushui Yifang Shan (Half a Pond of Autumn Water and a House of Mountain).

Bibliography

[1] FENG, G. (1974) 'Xianzhi Tang Gao' (Manuscripts from Xian-Zhi Tang) in Shen, Y. (ed) *Zhong Guo Jin Dai Shi Liao* (Historical Data from Post-1840 China). Taipei: Wen Hai Press (in traditional Chinese).

[2] LI, M. & FENG, G. (1991) *Tongzhi Suzhou Fu Zhi* (1862–1874 Compiled Records of Suzhou). Nanjing: Jiangsu Guji Press (in traditional Chinese).

[3] LIU, D. (2005) *Suzhou Gudian Yuan Lin* (Classical Gardens of Suzhou). Beijing: China Architecture & Building Press (in simplified Chinese).

[4] JIN, T. (2008) 'Yiyuan Ji' (on the Garden of Recuperation) in Wang, J. (ed) *Suzhou Yuan Lin Li Dai Wen Chao* (Selected Historical Literature on Classical Gardens of Suzhou). Shanghai: Shanghai Jointed Book Store (in simplified Chinese).

Credits

Surveyors:

2014 Undergraduate Class of Landscape Architecture (in groups):

ZHENG Xujing, YANG Xu, SUN bingyu, LEI Kai, LI Ruoyu, GAO Yifan, LIANG Ge, JIA Wenjing; LI Guoqing, ZHONG Huimin, SHEN Erdi, ZHANG Wenxin, ZHANG Kehua, HU Jia; GUO Xiaonan, LI Miaomiao, MA Qian, MA Yu, WANG Wei, DONG Qiwei, XU Tingting; XUE Yuan, XU Shiwen, YANG Fei, YANG Yujing, DING Wanjing, LUO Weizhen, ZHANG Wenjie.

Tutors: LIN Yuan and YUE Yanmin

Drawing Preparation: WEN Wujuan, WANG Yibo, YANG Xuanzi, WEI Qi

Date: October, 2012

图5：主山与问泉亭
Fig.5: Primary rockery hill and Wenquan Ting

图6：自问泉亭看四面厅
Fig.6: A view of Simian Ting from Wenquan Ting

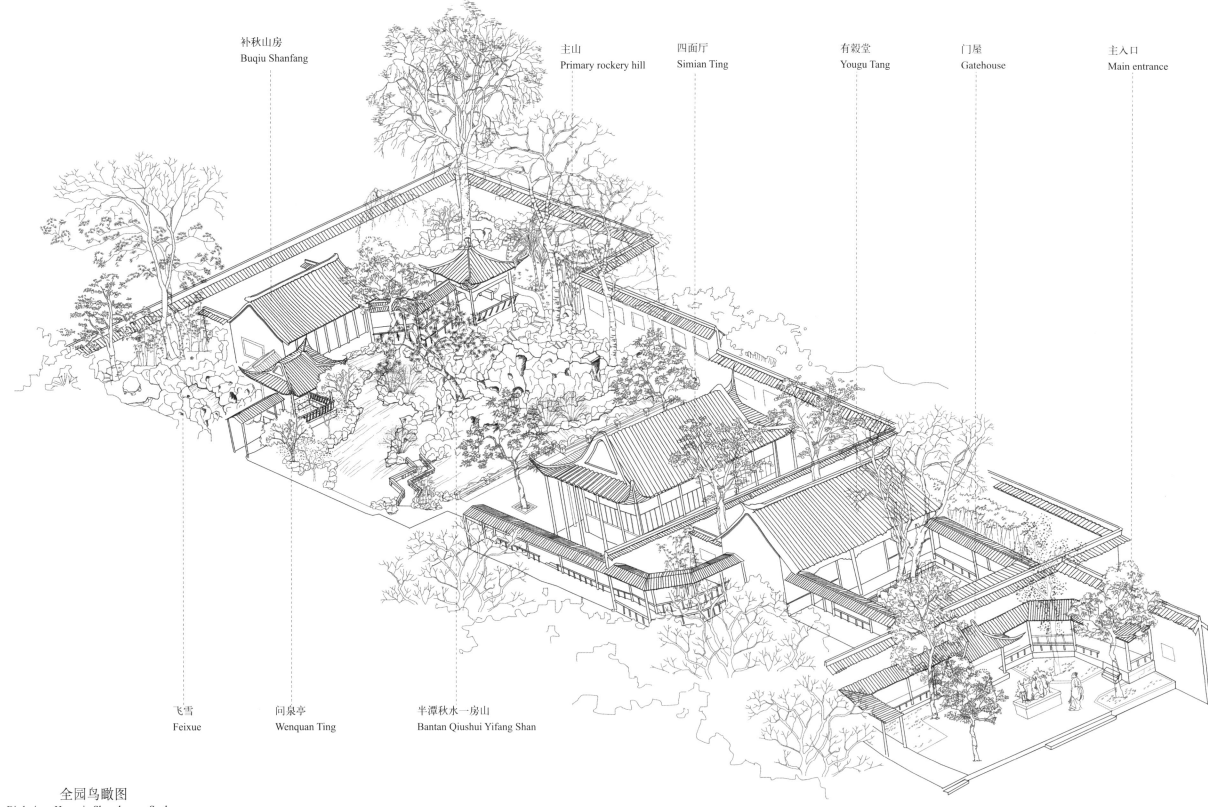

全园鸟瞰图
Bird-view: Huanxiu Shanzhuang, Suzhou

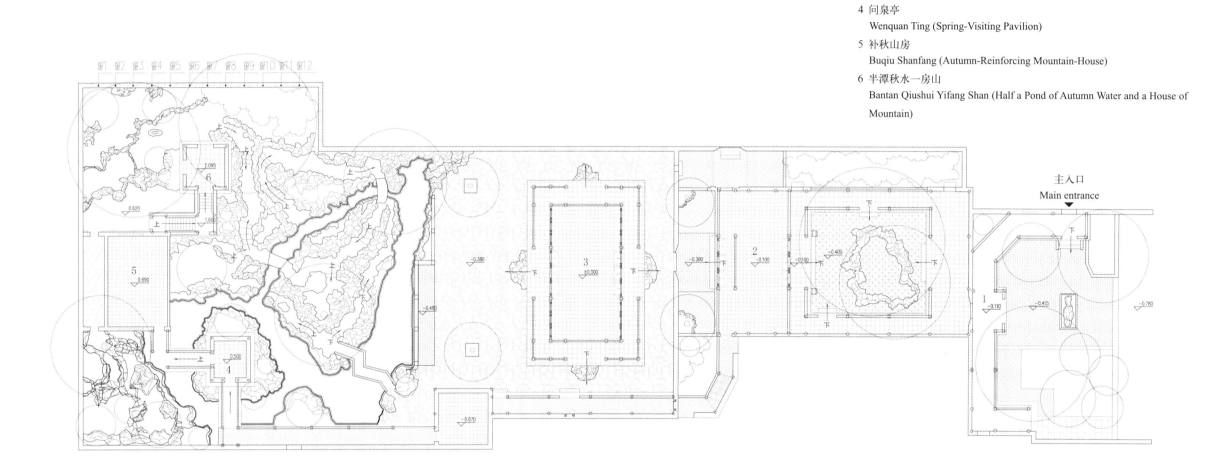

1 门屋
  Gatehouse
2 有榖堂
  Yougu Tang (Good Harvest Hall)
3 四面厅
  Simian Ting (Quadrilateral Opening Hall)
4 问泉亭
  Wenquan Ting (Spring-Visiting Pavilion)
5 补秋山房
  Buqiu Shanfang (Autumn-Reinforcing Mountain-House)
6 半潭秋水一房山
  Bantan Qiushui Yifang Shan (Half a Pond of Autumn Water and a House of Mountain)

总平面图
Site plan: Ground level

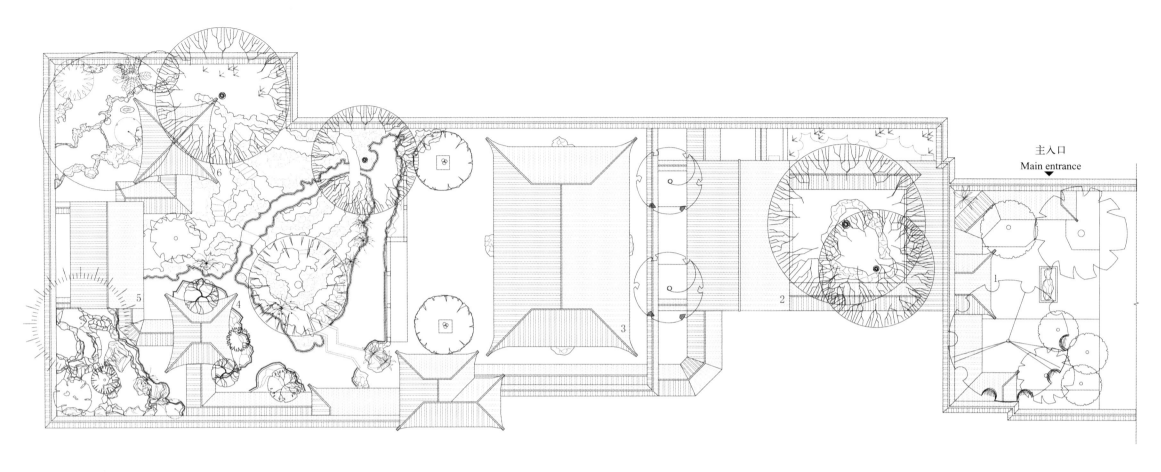

植物配置表
Vegetation

植物图例 Legend	植物名称 Binomial	植物科属 Family
	桂花树 Osmanthus fragrans	木犀科木犀属
	白皮松 Pinus bungeana	松科松属
	油松 Pinus tabulaeformis	松科松属
	瓜子黄杨 Buxus sinica	黄杨科黄杨属
	糙叶树 Aphananthe aspera	榆科糙叶树属
	朴树 Celtis sinensis	榆科朴属
	白玉兰 Magnolia denudata	木兰科木兰属
	银杏 Ginkgo biloba	银杏科银杏属
	山茶 Camellia japonica	山茶科山茶属
	石榴 Punica granatum	石榴科石榴属
	鸡爪槭 Acer palmatum	槭树科槭树属
	五角枫 Acer pictum subsp. mono	槭树科槭树属
	紫薇 Lagerstroemia indica	千屈菜科紫薇属
	桃叶珊瑚 Aucuba chinensis	山茱萸科桃叶珊瑚属
	碧桃 Amygdalus persica var. duplex	蔷薇科李属
	芭蕉 Musa basjoo	芭蕉科芭蕉属
	南天竹 Nandina domestica	小檗科南天竹属
	连翘 Forsythia suspensa	木犀科连翘属
	箬竹 Indocalamus tessellatus	禾本科箬竹属

1 门屋 Gatehouse
2 有穀堂 Yougu Tang (Good Harvest Hall)
3 四面厅 Simian Ting (Quadrilateral Opening Hall)
4 问泉亭 Wenquan Ting (Spring-Visiting Pavilion)
5 补秋山房 Buqiu Shanfang (Autumn Reinforcing Mountain-House)
6 半潭秋水一房山 Bantan Qiushui Yifang Shan (Half a Pond of Autumn Water and a House of Mountain)

屋顶平面图
Site plan: Roof level

东围墙漏窗 1-12 大样图
Details: *Louchuang* (traceried window openings) nos.1-12, east perimeter wall

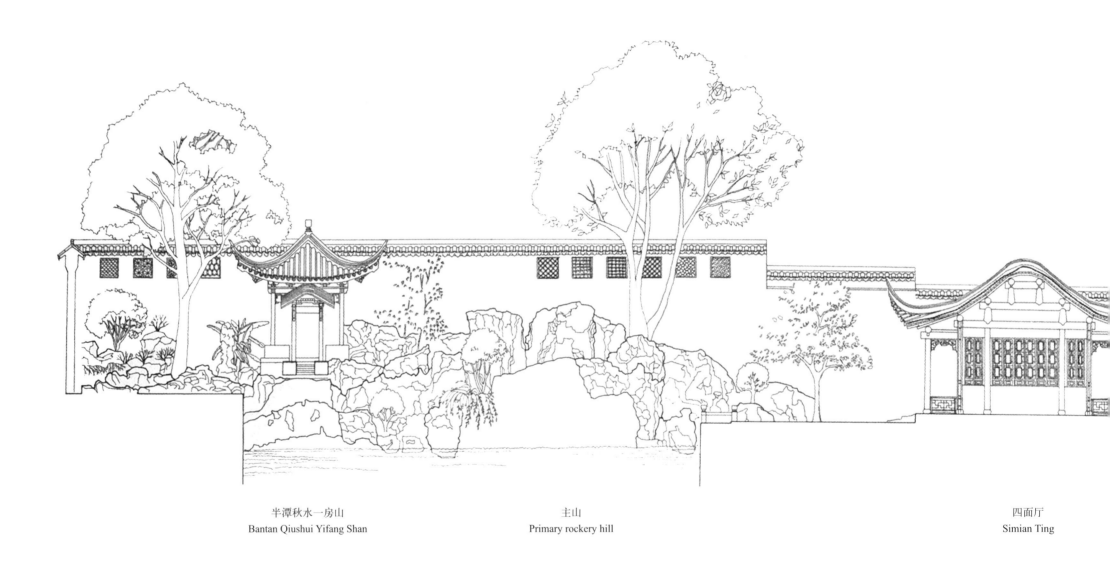

| 半潭秋水一房山 | 主山 | 四面厅 |
| Bantan Qiushui Yifang Shan | Primary rockery hill | Simian Ting |

全园南-北剖面图（东视）
North-south section: Huanxiu Shanzhuang

有谷堂　　　　　　　　　门屋　　　　　　　主入口
Yougu Tang　　　　　　　Gatehouse　　　　Main entrance

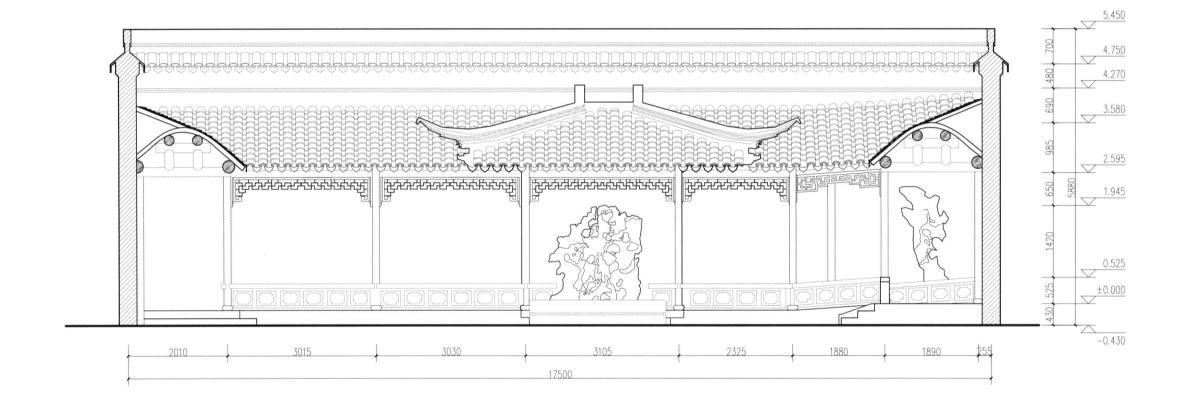

门屋南立面图
South elevation: Gatehouse

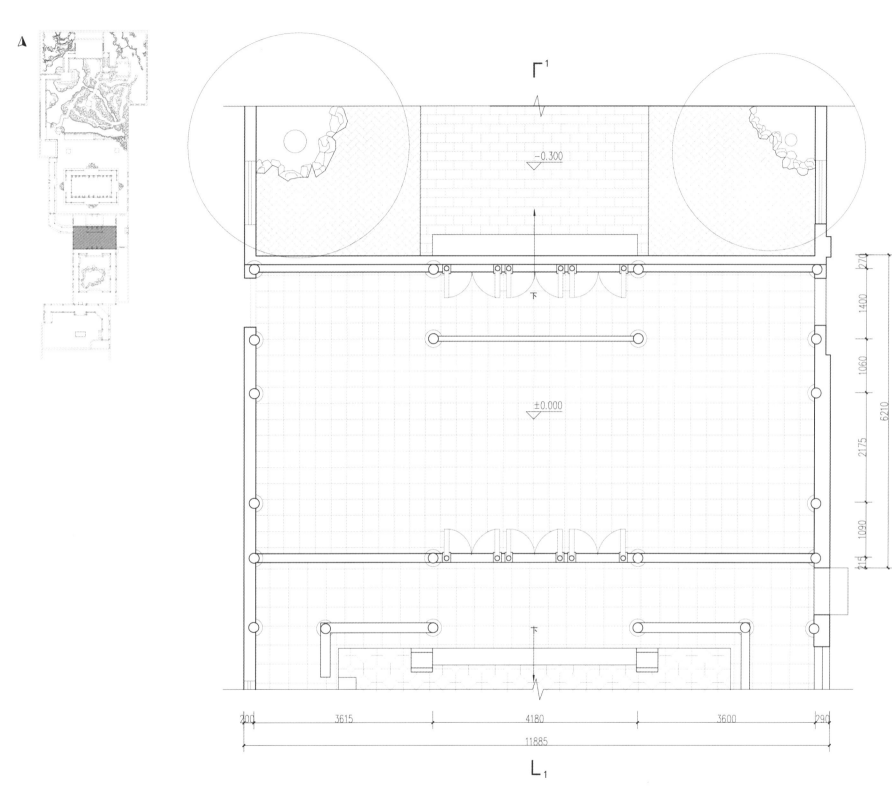

有穀堂平面图
Plan: Yougu Tang

有穀堂室外
Yougu Tang: Courtyard

有穀堂室内
Yougu Tang: indoors

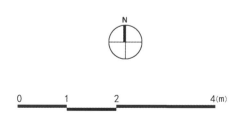

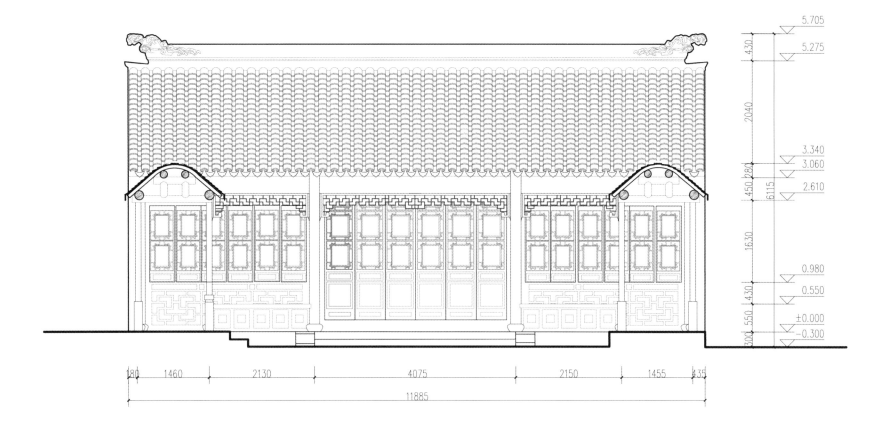

有穀堂南立面图
South elevation: Yougu Tang

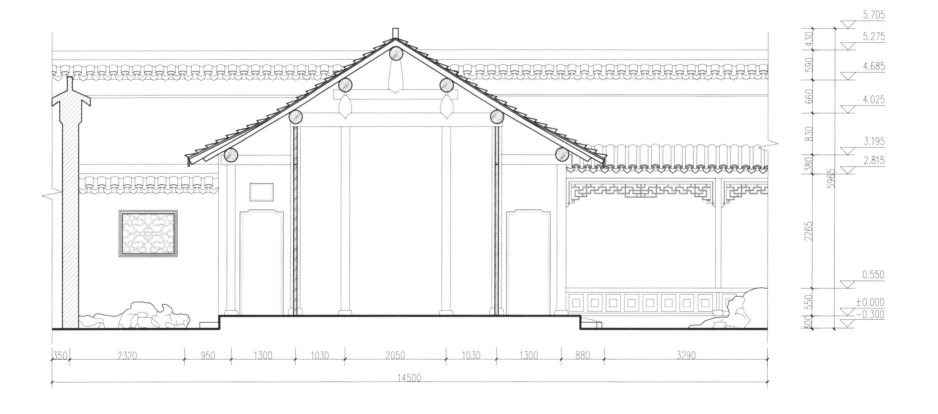

有榖堂 1-1 剖面图
Section 1-1: Yougu Tang

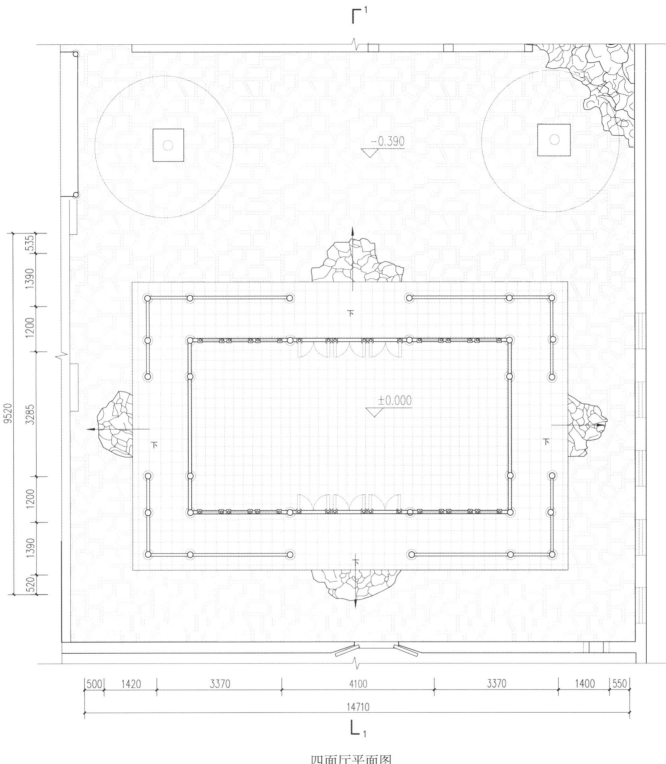

四面厅
Simian Ting

四面厅平面图
Plan: Simian Ting

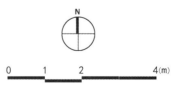

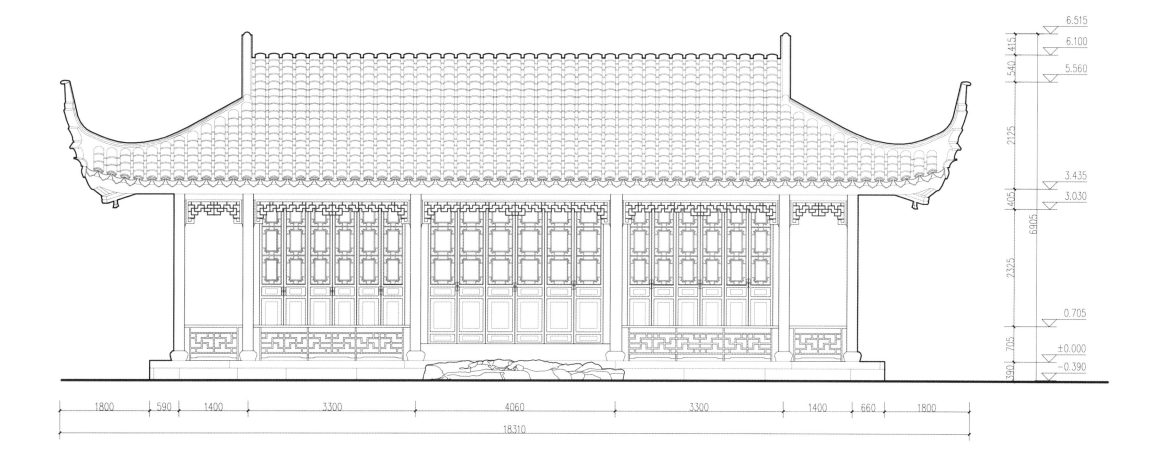

四面厅北立面图

North elevation: Simian Ting

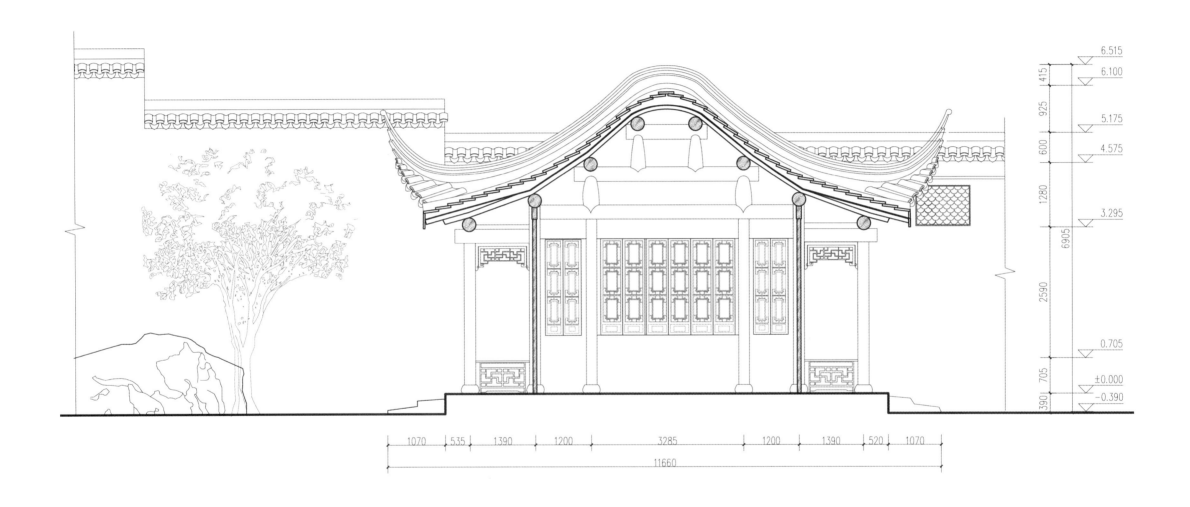

四面厅 1-1 剖面图
Section1-1: Simian Ting

问泉亭
Wenquan Ting

补秋山房—半潭秋水一房山
Buqiu Shanfang and Bantan Qiushui Yifang Shan

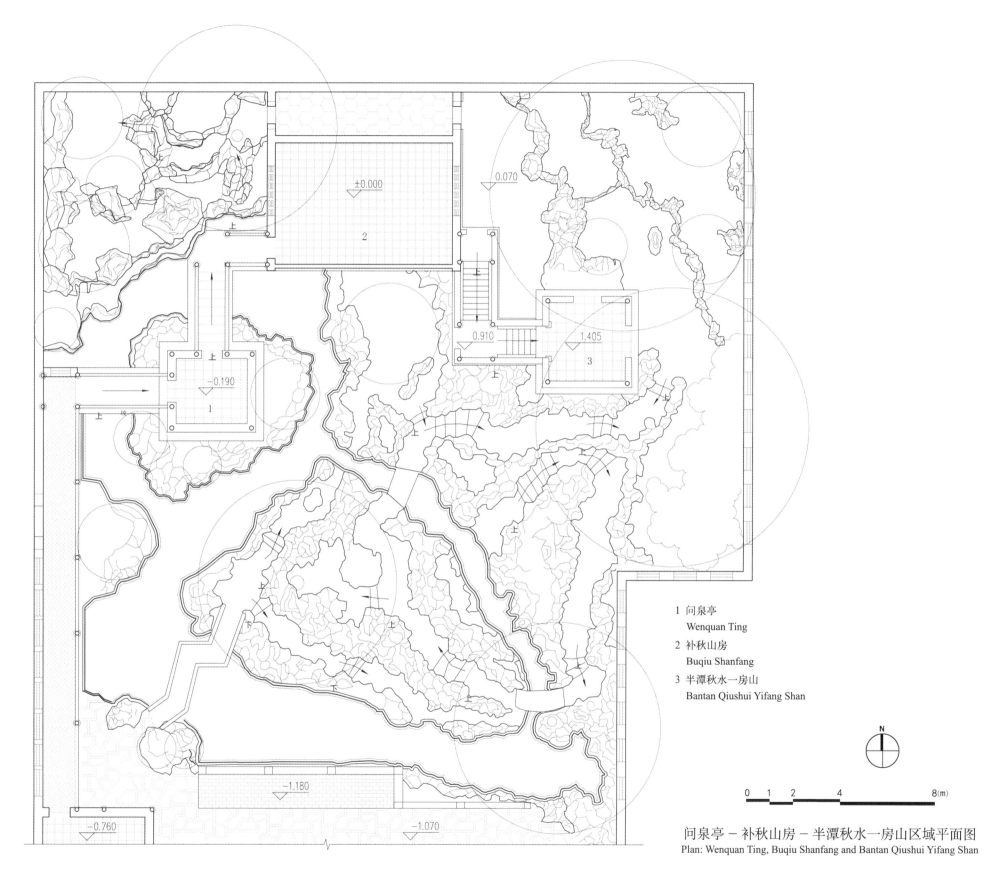

1 问泉亭
　Wenquan Ting
2 补秋山房
　Buqiu Shanfang
3 半潭秋水一房山
　Bantan Qiushui Yifang Shan

问泉亭－补秋山房－半潭秋水一房山区域平面图
Plan: Wenquan Ting, Buqiu Shanfang and Bantan Qiushui Yifang Shan

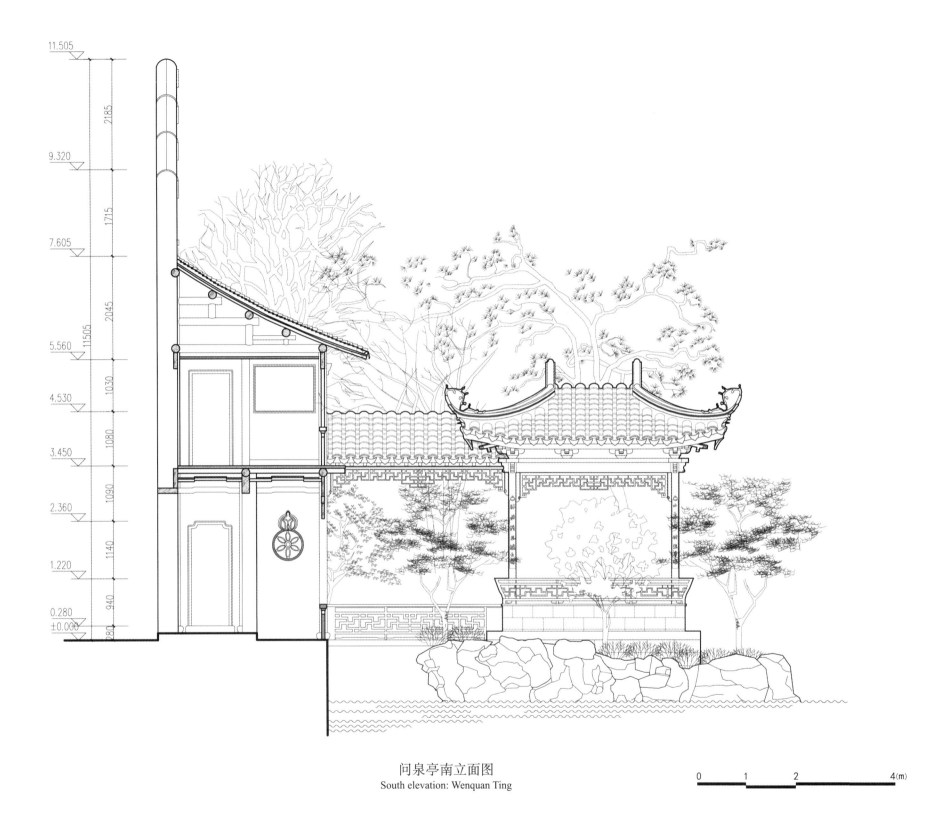

问泉亭南立面图
South elevation: Wenquan Ting

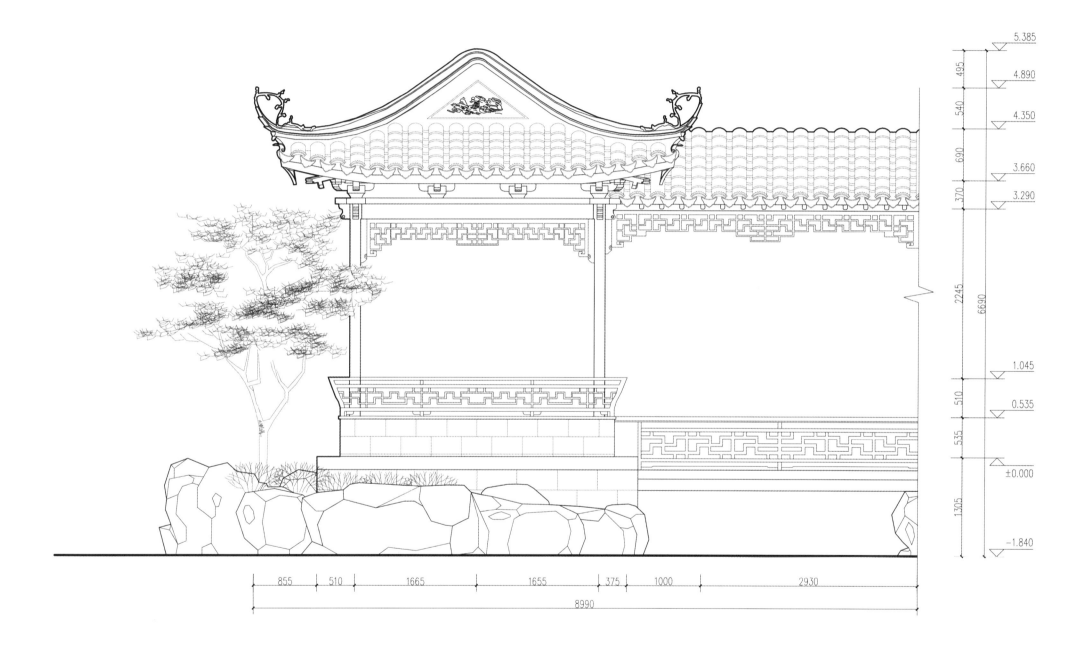

问泉亭东立面图
East elevation: Wenquan Ting

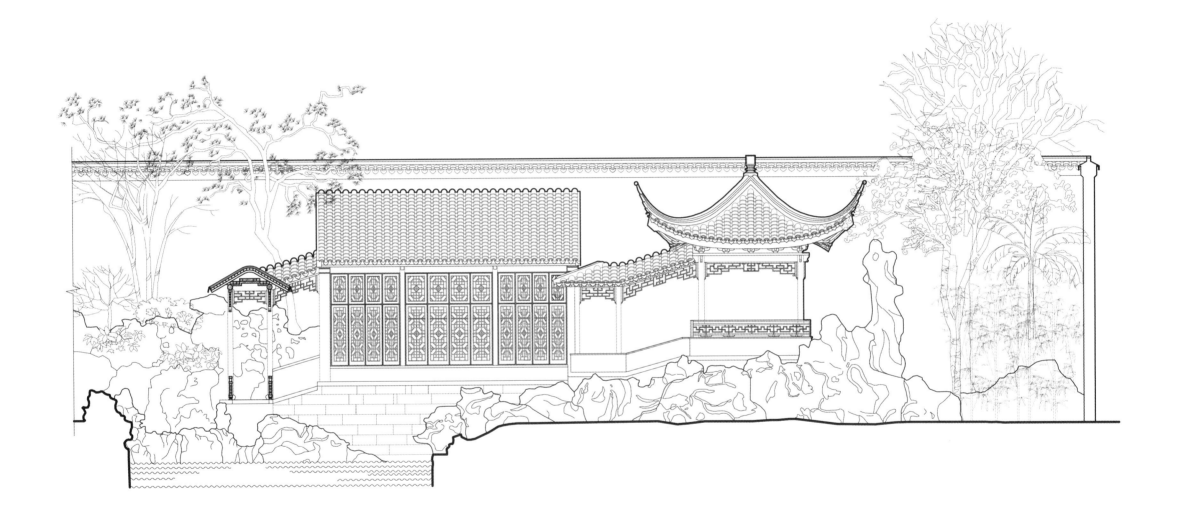

补秋山房—半潭秋水一房山南立面图
South elevation: Buqiu Shanfang and Bantan Qiushui Yifang Shan

西长廊东立面图
East elevation: West galleries

"飞雪"假山南立面图
South elevation: Feixue, Secondary rockery hill

"飞雪"假山
Feixue

主山北立面图
North elevation: Primary rockery hill

主山西立面图
West elevation: Primary rockery hill

苏州耦园

Ou Yuan (Couple's Garden Retreat), Suzhou

· 地址：苏州市仓街小新桥巷
· 年代：清
· 保护级别：世界文化遗产，全国重点文物保护单位（第五批）

· Address: 6 Xiaoqinqiao Ln., Cang St., Gusu Dist., Suzhou
· Origin: 17th Century
· Status: UNESCO World Cultural Heritage Site (2000); Key Cultural Heritage Site under State Protection (2001)

03 苏州耦园

耦园位于苏州老城东北部仓街以东的小新桥巷，东面紧邻东城墙与内城河，较东城墙北门楼门稍南；南面开正门，门前为水巷；北面也是河道，故全园南、北、东三面均临河，仅西面临街。全园占地约12亩，宅院居中，园子在宅的东西两侧，习称为东花园和西花园。清初，保宁（今属四川阆中）知府陆锦在此处筑宅，宅东辟园，名为"涉园"，此即是耦园东花园的前身。陆氏之后，园为崇明（今属上海市）祝氏所得。清光绪年间（1875—1908年），安徽巡抚沈秉成辞官后落户苏州，得到涉园，于是进行了整筑并新建了西花园，形成了两园夹一宅的格局并保持至今，园子也更名为耦园。沈秉成卒后，其后人返回故籍，耦园遂交与他人代管，后陆续出租为民房，逐渐败落。民国29年（1940年），常州实业家刘国钧买下了耦园，开始进行修复；1950年，中路宅院部分的主厅堂（今之载酒堂）失火被毁；1956年，耦园归振亚丝织厂管理，作为车间、仓库、职工宿舍以及托儿所使用；1960年，耦园归苏州市园林管理处管理，得到了全面修复，由刘敦桢、陈从周二位先生担任修复工程的现场指导。1995年耦园列为江苏省文物保护单位，2001年列为第五批全国重点文物保护单位。2000年，耦园作为"苏州古典园林"的扩展项目列入《世界遗产名录》。

耦园宅院居中，由门厅、主厅载酒堂、楼大厅等建筑组成三进院落。东园以黄石假山居中，各主要建筑环绕假山布列；山北为主厅城曲草堂，山东为长廊及吾爱亭；山南临池，池南岸为水榭"山水间"，再南为听橹楼、魁星楼与长廊；山西为藤花舫、长廊与无俗韵轩小院。西园以织帘老屋及屋前假山为主体，山南为纫兰室一组院落；织帘老屋以东是鹤寿亭、以北是藏书楼。

图1："厚德载福"墙门
Fig.1: Gate bearing the inscription 'Houde Zaifu' (Virtue Brings Fortune)

主要参考文献：
[1] 周退密. 重修耦园记 [Z]. 吴湊题. （碑刻）.
[2] 曹允源、李根源. 民国吴县志 [M]. 中国地方志集成·江苏府县志辑. 第十一至十二册. 据民国二十二年苏州文新公司铅印本影印. 八十卷. 江苏古籍出版社，1991.
[3] 苏州市园林和绿化管理局. 耦园志 [M]. 上海：文汇出版社，2013.

测　　绘：西安建筑科技大学建筑学院
　　　　　景观建筑学2010级—
　　　　　高洁、付梦晗、高义、李孟军、刘嘉伟、王祯、汶武娟、徐传语、崔文睿、戴梦蓉、王乙惠、赵杰、郭建廷、贾川、李霄、翁婧雯、许菲菲、张佳琪、张元凯、汪科磊、裴宁、王旭红、王樱子、史敏慎、关键、江畅、慈硕文、何田、李绍伦、文娟；
　　　　　建筑历史与理论研究生2013级—
　　　　　申佩玉、黄思达、夏楠、冯卫杰、申田野、郭润泽、王泽暄、赵泽群、马晓鸣、朱立波；
　　　　　建筑学2010级01班—
　　　　　黄晓娜、江曼、李菁华、万少帅、王涛、于涛、何建、高迎衔、刘意杰、万素影、王霞、林思佳、刘妮、赵子良、张书苑、杨梅、康路、刘洋；范诗琪、郭芸暄、石嘉怡、张琪、张嵩；
指导教师：林源、岳岩敏；
图纸绘制、整理：杨桓子、夏楠；
测绘时间：2013年10月

03. Ou Yuan (Couple's Garden Retreat), Suzhou

The garden is located at the Cang Street and Xiaoxinqiao Lane crossing in the northeastern part of the historic urban area of Suzhou. It abuts the city's inner moat and wall at its east and not far from the northernmost east city gate to its north. With the city moat and river courses running at its north and south, the site is surrounded by water on three sides, leaving its west side on Cang Street. The garden covers approx. 8,000 m² of land which has the 'housing part' in the middle flanked at both sides by gardens known as the 'East Garden' and 'West Garden'. In early Qing Dynasty (1644–1912), Lu Jin, the former prefect of Baoning (present-day Langzhong, Sichuan Prov.), built his residence here and laid down a garden, named She Yuan (Garden of Strolling), which would later become the East Garden. She Yuan later came into the possession of a Zhu family from Chongming (present-day Chongming, Shanghai), and during Emperor Dezong's reign (1875–1908) was purchased by Shen Bingcheng (1823–1895), the retired prefect of Anhui Province. Shen's contribution to the site included not only some repair work, but as well an addition of the West Garden, which finally established the layout, along with the name, we know today. After Shen's death, his family returned to their home of origin, leaving the Garden at the mercy of caretakers who

图 2：载酒堂
Fig.2: Zaijiu Tang

图 4：东园樨廊
Fig.4: Xi Lang, East Garden

图 3："诗酒联欢" 墙门
Fig.3: Gate bearing the inscription 'Shijiu Lianhuan' (Musenal and Bacchanal Joy)

图 5：吾爱亭
Fig.5: Wu'ai Ting

图 6：山水间
Fig.6: Shanshui Jian

图 7：听橹楼
Fig.7: Tinglu Lou

allowed it to be let to various tenants and gradually fall into decline. In 1940, the already dilapidated property was acquired and also restored to a certain extent by LIU Guojun (1887–1978), only to be ravaged again ten years later by an accidentally caused fire, which razed its main hall (present-day Zaijiu Tang). In 1956 the site found its way among the properties owned by Zhenya Textile Mill and was adaptively reused as workshops, storages, employee dormitories and a kindergarten. Finally the garden's vicissitudinous fate was secured after being put under the care of the Suzhou Municipal Bureau of Gardens, which commissioned extensive restorations under the instruction and supervision of the renowned architectural historians Prof. Liu Dunzhen (1897–1968) and Prof. Chen Congzhou (1918–2000). In 1995, the site was listed as a Cultural Heritage Site under Provincial Protection and in 2001 a Key Cultural Heritage Site under State Protection (5th Round of Nomination). It was also recognized by UNESCO World Heritage Center, which included it, as part of extension, in the inscribed World Cultural Heritage Site *Classical Gardens of Suzhou* in 2000.

The 'housing part' of the Garden comprises of a succession of 3 courtyard complexes including those of the Gatehouse, Zaijiu Tang (Wine-Carrying Hall), the main hall, and Lou Ting (Two-Storied Hall). The East Garden is dominated by the central rockery hill constructed with yellowish stones, around which its garden structures are arranged. These structures include the hill's north-side main hall Chengqu Caotang (City-Corner Thatched Cottage), east-side gallery and Wu'ai Ting (Abode-Loving Pavilion), and west-side Tenghua Fang (Wisteria-Blossom Boat), Osmanthus Galleries and the court of Wu Suyun Xuan (Mundanity-Free Pavilion). The hill is also adjoined on its south by a pond, on the southern bank of which stands Shanshui Jian (Within Landscape), a waterside pavilion. Further south to the pavilion there are Tinglu Lou (Boat-Rowing Sound Building), Kuixing Ge (Literati-Star Tower) and galleries. The West Garden also has a rockery hill along with Zhilian Laowu (Curtain-Knitting Hall) at its north and another court featuring Renlan Shi (Orchid-Stitching House) on its south. From Zhilian Laowu one may ramble eastward to reach Heshou Ting (Longevity Pavilion) or northward to Cangshu Lou (Library).

Bibliography

[1] ZHOU, T. *Chong-Xiu Ou-Yuan Ji* (On the Restoration of Ou Yuan) (a tablet inscription), consulted Oct 01, 2016 (in traditional Chinese).

[2] CAO, Y. & Li, G. (1991) *Minguo Wuxian Zhi* (ROC Compiled Records for the County of Wu). Nanjing: Jiangsu Guji Press (in traditional Chinese).

[3] Suzhou Municipal Bureau for Gardening and Forestation. (2013) *Ou Yuan Zhi* (Records of the Couple's Garden Retreat. Shanghai: Wenhui Press (in Simplified Chinese).

Credits

Surveyors:

2015 Undergraduate Class of Landscape Architecture (in groups):

GAO Jie, FU Menghan, GAO Yi, LI Mengjun, LIU Jiawei, WANG Zhen, XU Chuanyu, CUI Wenrui, DAI Mengrong, WANG Yihui, WEN Wujuan, ZHAO Jie; GUO Jianting, JIA Chuan, LI Xiao, WENG Jingwen, XU Feifei, ZHANG Jiaqi, ZHANG Yuankai, WANG Kelei, PEI Ning, WANG Xuhong, WANG Yingzi, SHI Minshen; JIANG Chang, CI Shuowen, GUAN Jian, HE Tian, LI Shaolun, WEN Juan

2016 Postgraduate Class of Architecture History:

SHEN Peiyu, HUANG Sida, XIA Nan, FENG Weijie, SHEN Tianye, GUO Runze, WANG Zexuan, ZHAO Zequn, MA Xiaoming, ZHU Libo

2015 Undergraduate Class A of Architecture (in groups):

HUANG Xiaona, JIANG Man, LI Jinghua, WAN Shaoshuai, WANG Tao, YU Tao; HE Jian, GAO Yingxian, LIU Yijie, WAN Suying, WANG Xia, LIN Sijia; LIU Ni, ZHAO Ziliang, ZHANG Shuyuan, YANG Mei, KANG Lu, LIU Yang; FAN Shiqi, GUO Yunxuan, SHI Jiayi, ZHANG Qi, ZHANG Song

Tutors: LIN Yuan, YUE Yanmin

Drawing Preparation: YANG Xuanzi, XIA Nan

Date: October, 2013

植物配置表
Vegetation

植物图例 Legend	植物名称 Binomial	植物科属 Family	植物图例 Legend	植物名称 Binomial	植物科属 Family	植物图例 Legend	植物名称 Binomial	植物科属 Family	植物图例 Legend	植物名称 Binomial	植物科属 Family
	香樟 Cinnamomum camphora	樟科樟属		梧桐 Firmiana simplex	梧桐科梧桐属		石榴 Punica granatum	石榴科石榴属		竹林 Bambusoideae	禾本科竹属
	白皮松 Pinus bungeana	松科松属		银杏 Ginkgo biloba	银杏科银杏属		五角枫 Acer pictum subsp. mono	槭树科槭树属		平枝枸子 Cotoneaster horizontalis	蔷薇科枸子属
	桂花 Osmanthus fragrans	木犀科木犀属		香橼 Citrus medica	芸香科柑橘属		木香 Rosa banksiae	蔷薇科风毛菊属		箬竹 Indocalamus tessellatus	禾本科箬竹属
	圆柏 Juniperus chinensis	松柏科圆柏属		木瓜 Chaenomeles sinensis	蔷薇科木瓜属		垂柳 Salix babylonica	杨柳科柳属		罗汉松 Podocarpus macrophyllus	罗汉松科罗汉松属
	紫荆 Cercis chinensis	豆科紫荆属		西柚 Citrus paradisi	芸香科柑橘属		绣球 Hydrangea macrophylla	虎耳草科绣球属		碧桃 Amygdalus persica var. duplex	蔷薇科李属
	黑松 Pinus thunbergii	松科松属		梅花 Armeniaca mume	蔷薇科杏属		芭蕉 Musa basjoo	芭蕉科芭蕉属		红枫 Acer palmatum 'Atropurpureum'	槭树科槭树属
	油松 Pinus tabulaeformis	松科松属		榆叶梅 Amygdalus triloba	蔷薇科桃属		腊梅 Chimonanthus praecox	腊梅科腊梅属		无患子 Sapindus saponaria	无患子科无患子属
	女贞 Ligustrum lucidum	木犀科女贞属		重阳木 Bischofia polycarpa	大戟科秋枫属		南天竹 Nandina domestica	小檗科南天竹属		八角金盘 Fatsia japonica	五加科八角金盘属
	侧柏 Platycladus orientalis	松柏科侧柏属		国槐 Sophora japonica	豆科槐属		山茶 Camellia japonica	山茶科山茶属		牡丹 Paeonia suffruticosa	毛茛科芍药属
	枇杷 Eriobotrya japonica	蔷薇科枇杷属		榆树 Ulmus pumila	榆科榆属		瓜子黄杨 Buxus sinica	黄杨科黄杨属		紫藤 Wisteria sinensis	豆科紫藤属
	樱花 Cerasus yedoensis	蔷薇科樱属		垂丝海棠 Malus halliana	蔷薇科苹果属		海桐 Pittosporum tobira	海桐科海桐花属			
	朴树 Celtis sinensis	榆科朴属		紫薇 Lagerstroemia indica	千屈菜科紫薇属		枸骨 Ilex cornuta	冬青科冬青属			
	白玉兰 Magnolia denudata	木兰科玉兰属		鸡爪槭 Acer palmatum	槭树科槭树属		凌霄 Campsis grandiflora	紫葳科凌霄属			

1	门屋 Gatehouse
2	谐隐双山 Xieyin Shuangshan (Couple's Retreat Mountains Chamber)
3	载酒堂 Zaijiu Tang (Wine-Carrying Hall)
4	楼厅 Lou Ting (Two-Storied Hall)
5	无俗韵轩 Wu Suyun Xuan (Mundanity-Free Pavilion)
6	榉廊 Xi Lang (Osmanthus Galleries)
7	藤花舫 Tenghua Fang (Wisteria-Bloossom Boat)
8	储香馆 Chuxiang Guan (Flagrance-Storing House)
9	城曲草堂 Chengqu Caotang (City-Corner Thatched Cottage)
10	还砚斋 Huanyan Zhai (Inkstone-Retrieving Study)
11	望月亭 Wangyue Ting (Moon-Viewing Pavilion)
12	吾爱亭 Wu'ai Ting (Abode-Loving Pavilion)
13	山水间 Shanshui Jian (Within Landscape)
14	听橹楼 Tinglu Lou (Boat-Rowing Sound Building)
15	魁星阁 Kuixing Ge (Literati-Star Tower)
16	藏书楼 Cangshu Lou (Library)
17	织帘老屋 Zhilian Laowu (Curtain-Knitting Hall)
18	鹤寿亭 Heshou Ting (Longevity Pavilion)
19	纫兰室 Renlan Shi (Orchid-Stitching Chamber)

总平面图
Site plan: Ground level

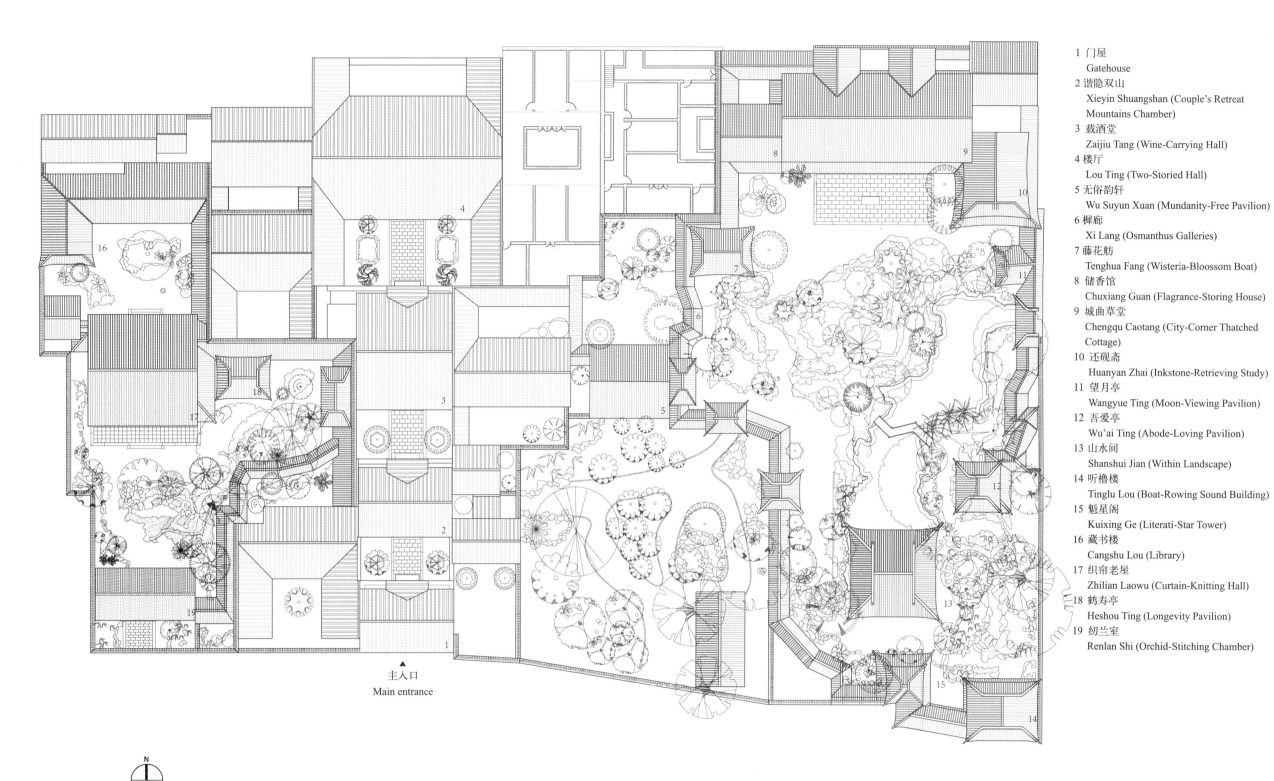

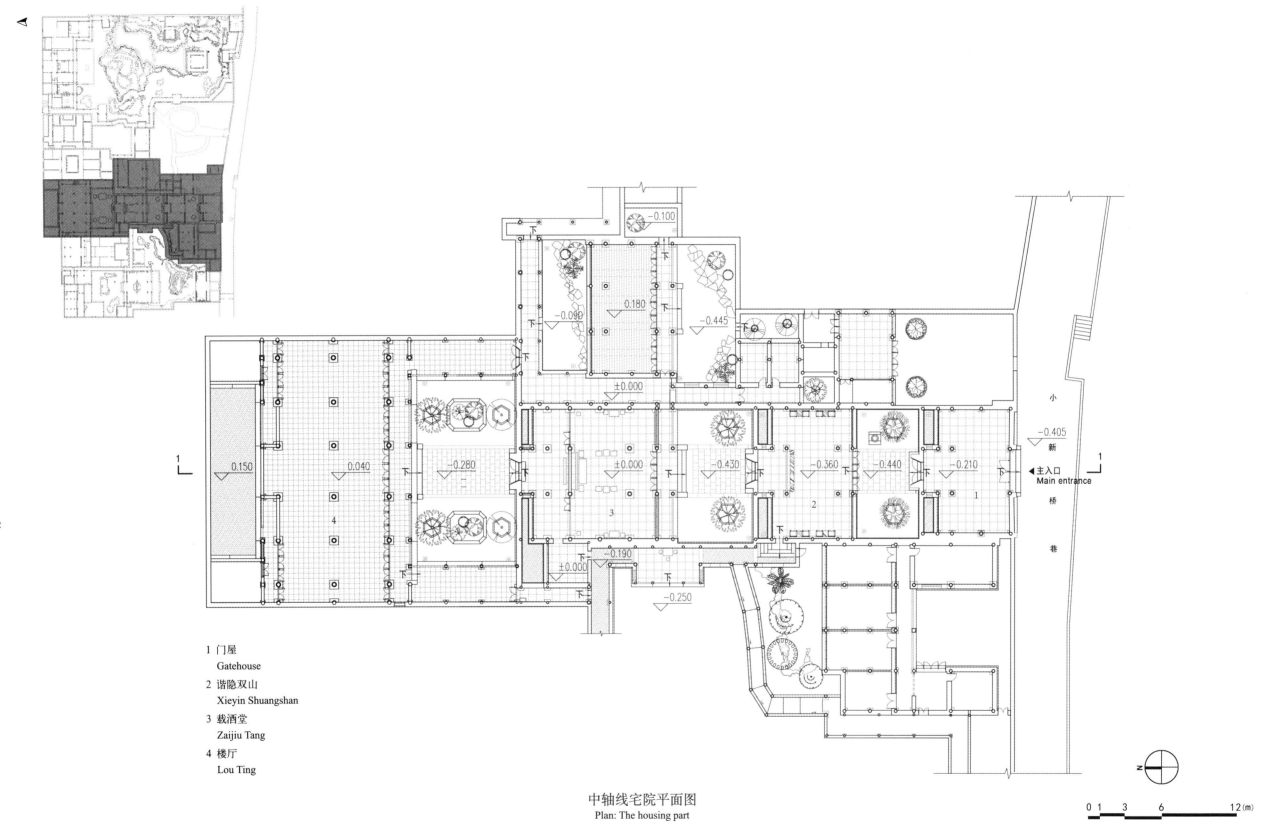

1 门屋
  Gatehouse
2 谐隐双山
  Xieyin Shuangshan
3 载酒堂
  Zaijiu Tang
4 楼厅
  Lou Ting

中轴线宅院平面图
Plan: The housing part

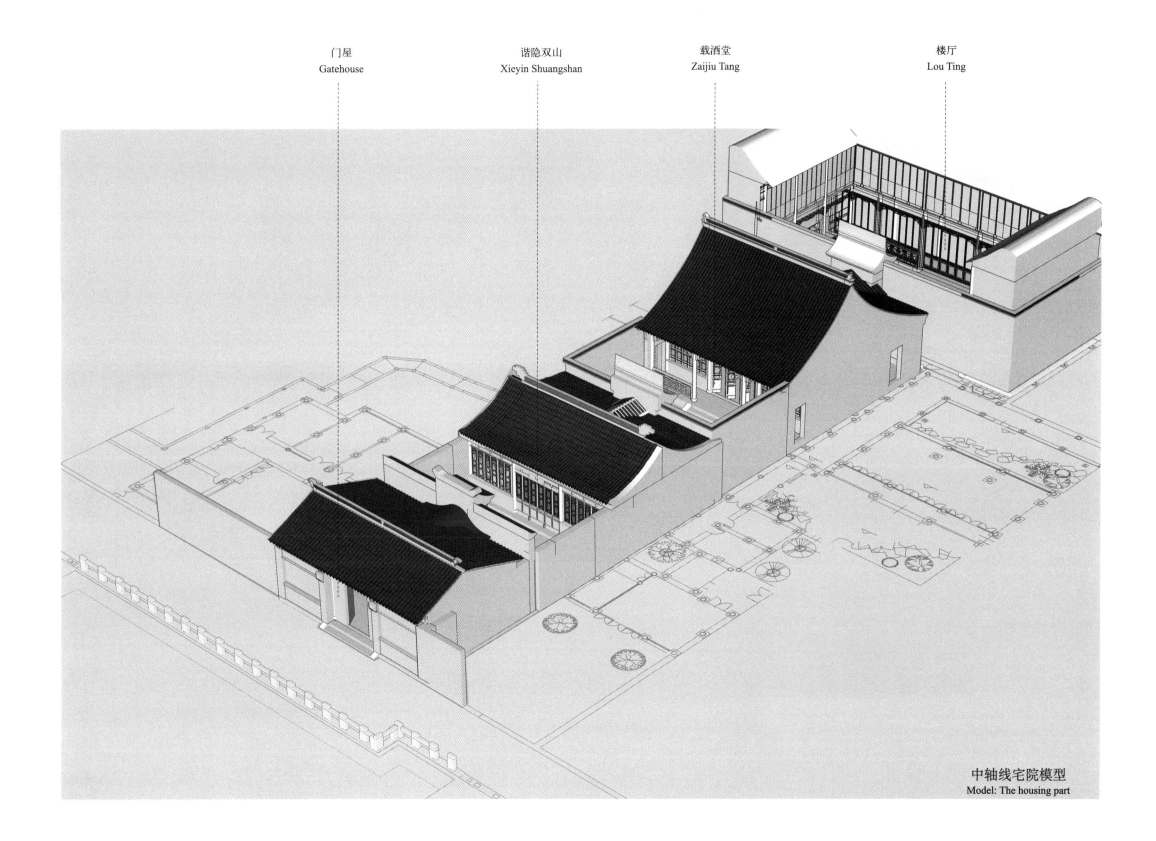

中轴线宅院模型
Model: The housing part

门屋 Gatehouse
谐隐双山 Xieyin Shuangshan
载酒堂 Zaijiu Tang
楼厅 Lou Ting

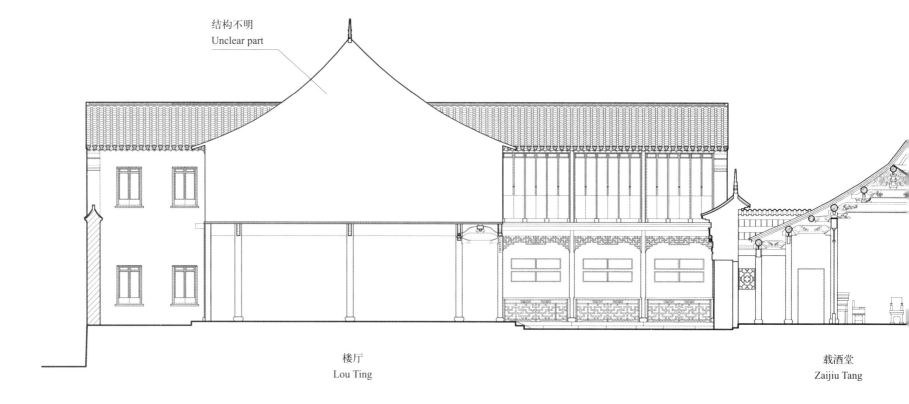

中轴线宅院 1-1 剖面图
Section 1-1: The housing part

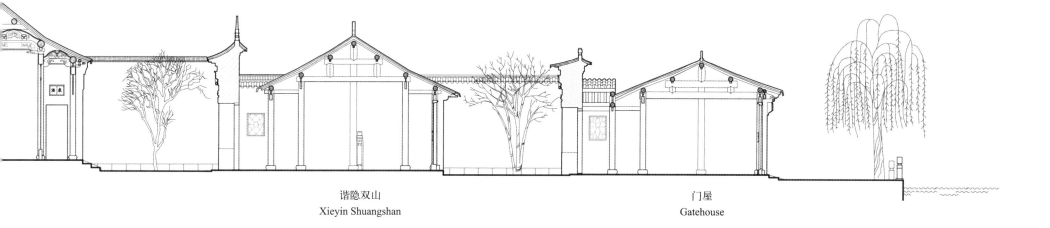

| 谐隐双山 | 门屋 |
| Xieyin Shuangshan | Gatehouse |

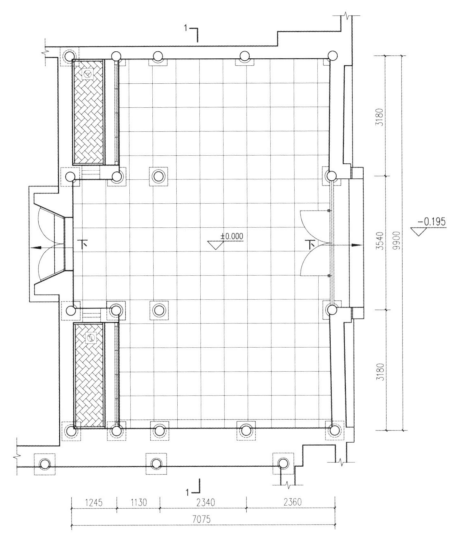

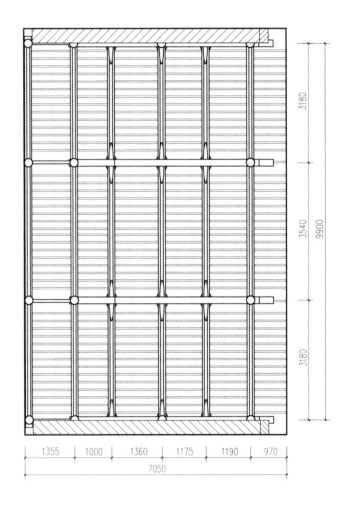

门屋平面图
Plan: Gatehouse

门屋梁架仰视图
Reflected truss plan: Gatehouse

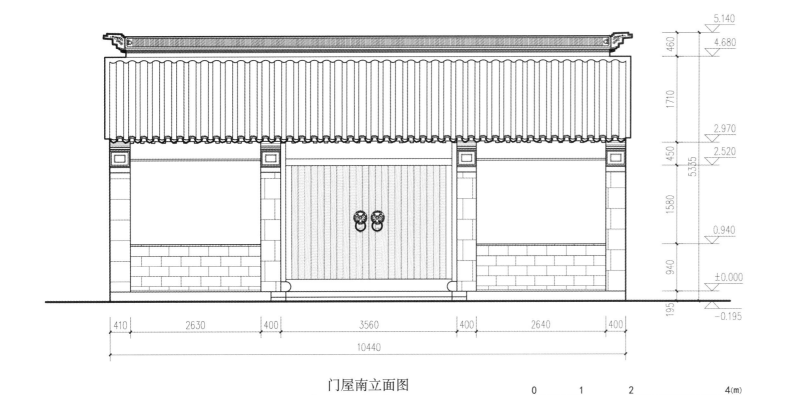

门屋南立面图
South elevation: Gatehouse

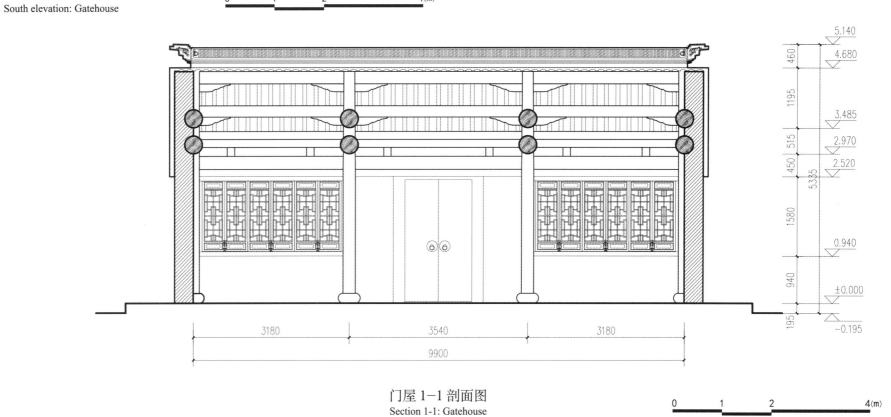

门屋 1-1 剖面图
Section 1-1: Gatehouse

"平泉小隐"墙门
Gate bearing the inscription 'Pingquan Xiaoyin'

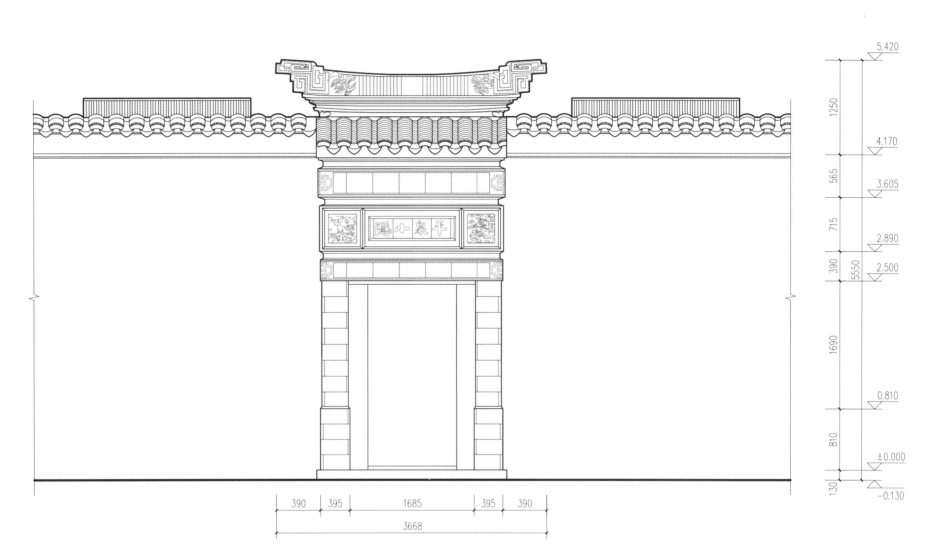

"平泉小隐"墙门北立面图
North elevation: Gate bearing the inscription 'Pingquan Xiaoyin' (Humble Retreat)

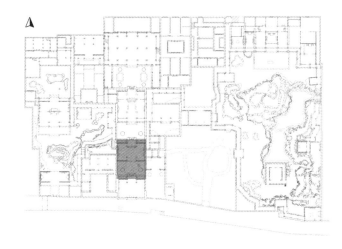

谐隐双山院落
Xieyin Shuangshan: courtyard

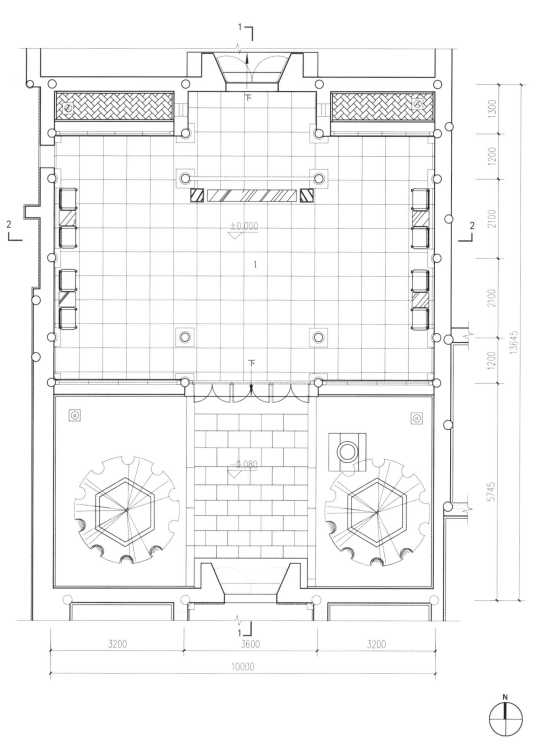

谐隐双山院落平面图
Plan: Xieyin Shuangshan

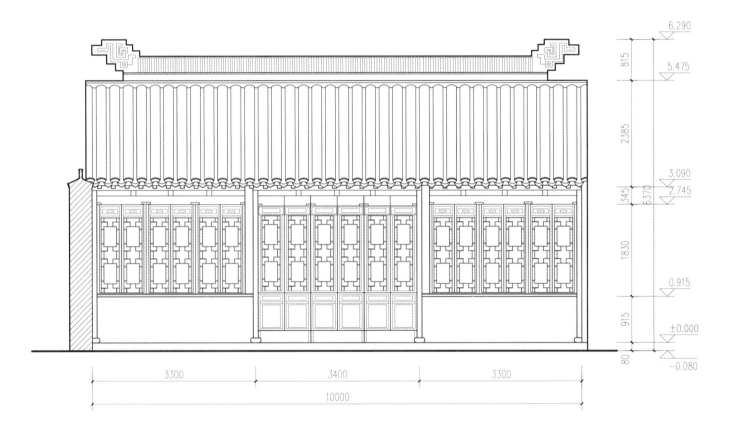

谐隐双山南立面图
South elevation: Xieyin Shuangshan

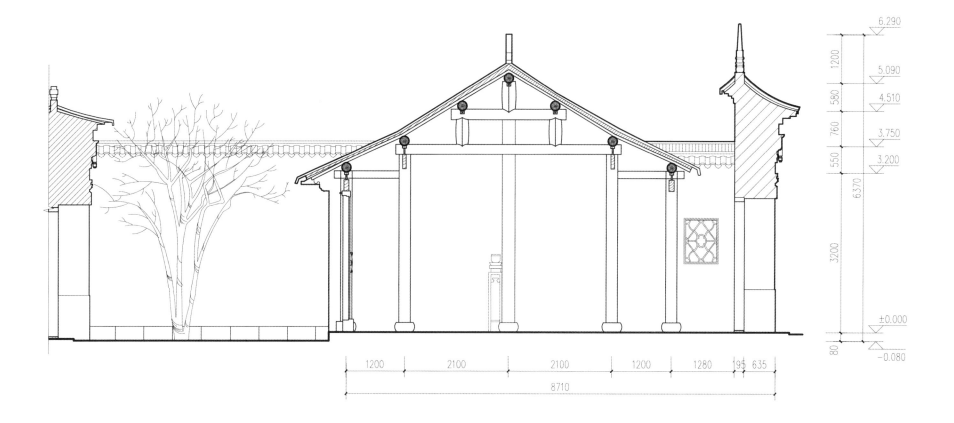

谐隐双山 1-1 剖面图
Section 1-1: Xieyin Shuangshan

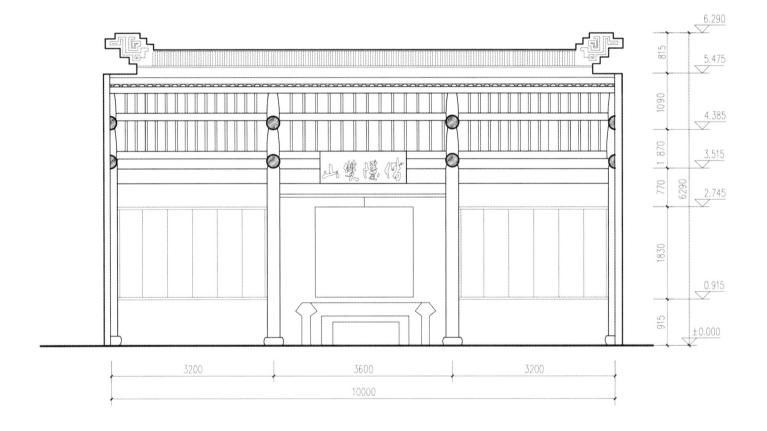

谐隐双山 2-2 剖面图
Section 2-2: Xieyin Shuangshan

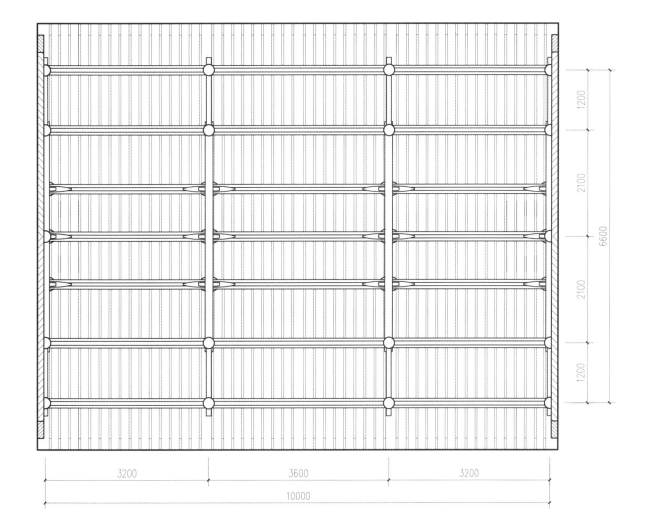

谐隐双山梁架仰视图
Reflected truss plan: Xieyin Shuangshan

谐隐双山室内
Xieyin Shuangshan: indoors

"厚德载福" 墙门
Gate bearing the inscription 'Houde Zaifu'

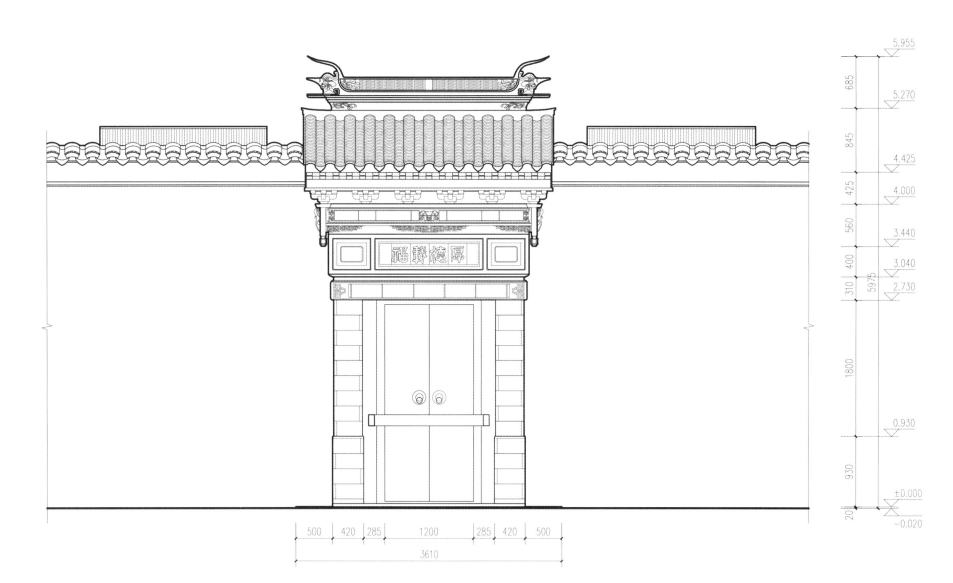

"厚德载福" 墙门北立面图
North elevation: Gate bearing the inscription 'Houde Zaifu' (Virtue Brings Fortune)

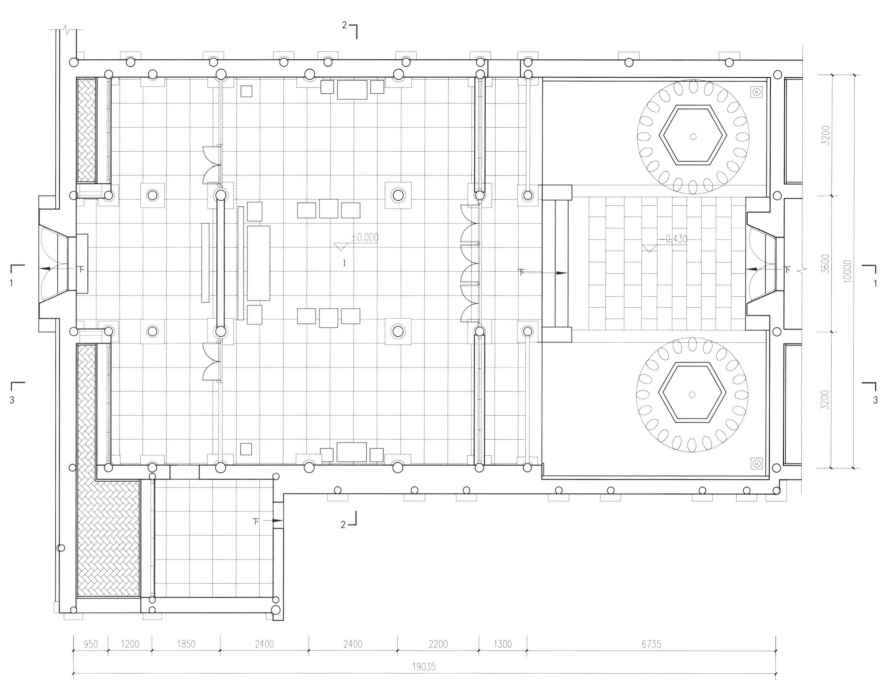

载酒堂院落平面图
Plan: Zaijiu Tang

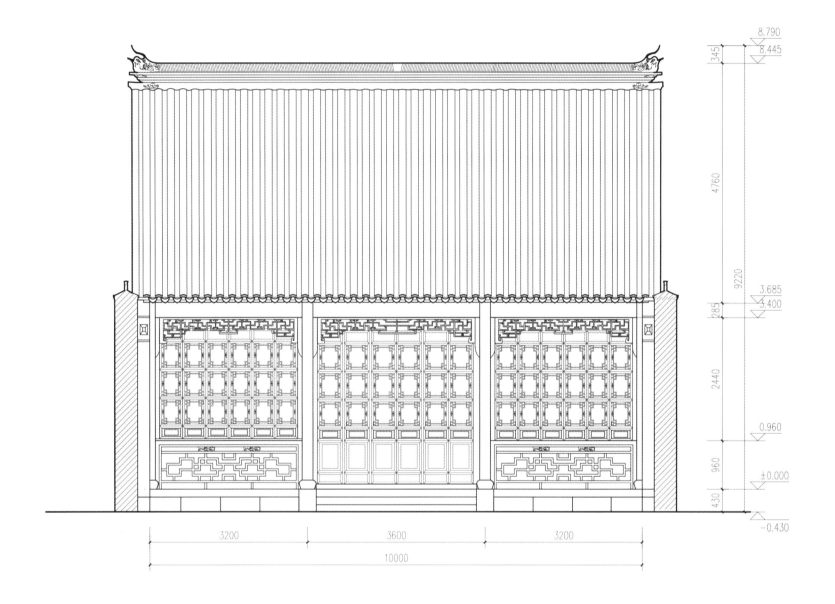

载酒堂南立面图
South elevation: Zaijiu Tang

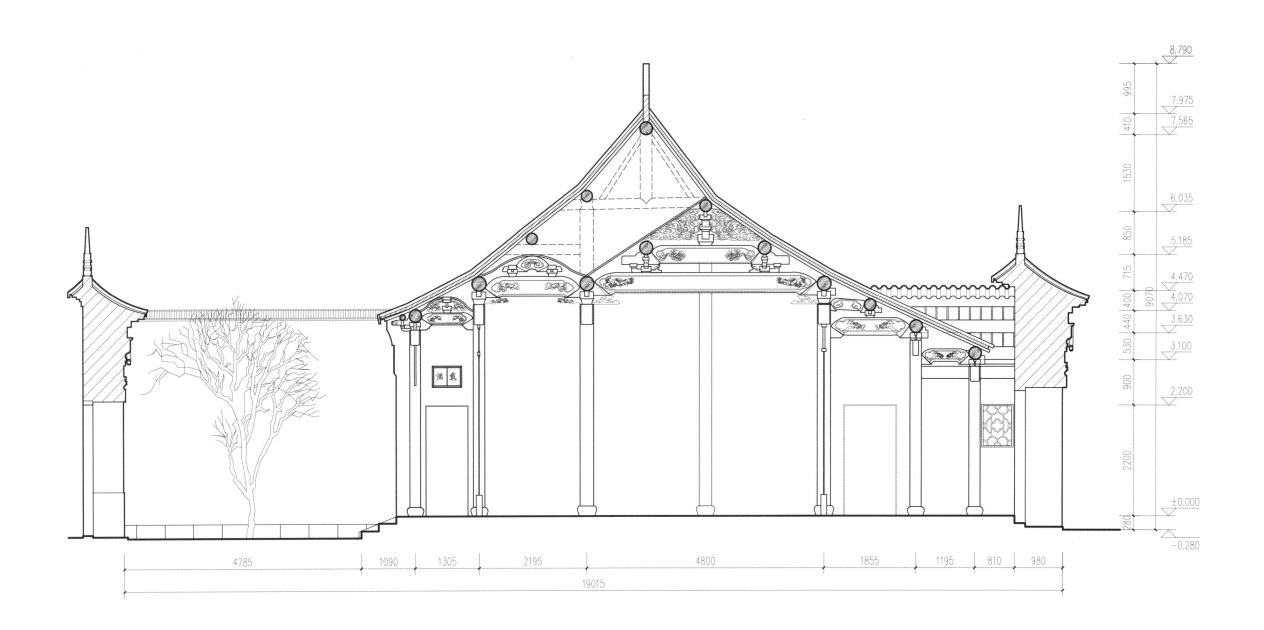

载酒堂院落 1-1 剖面图
Section 1-1: Zaijiu Tang

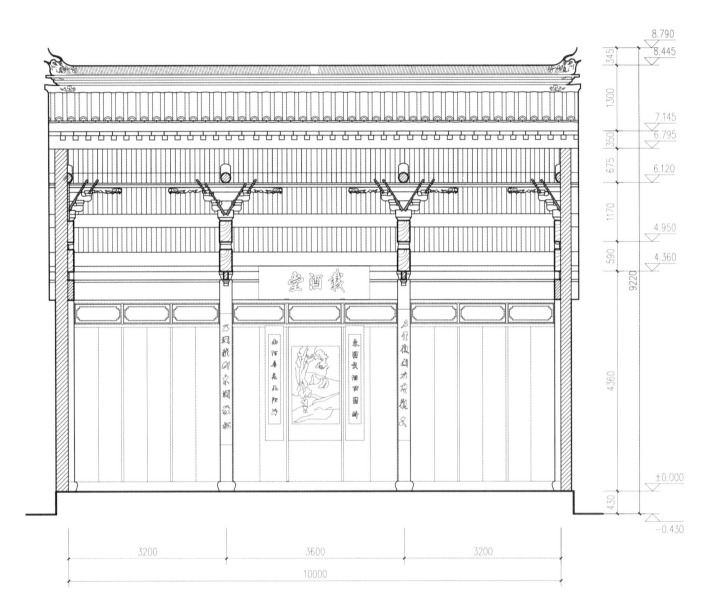

载酒堂 2-2 剖面图
Section 2-2: Zaijiu Tang

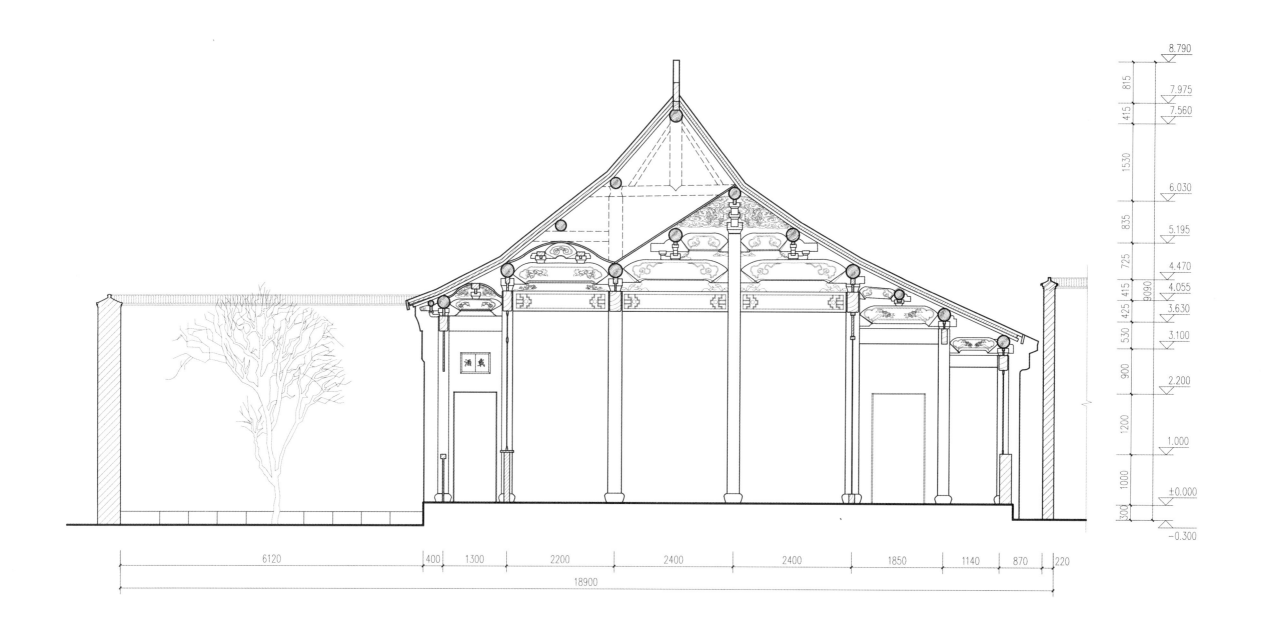

载酒堂院落 3-3 剖面图
Section 3-3: ZaijiuTang

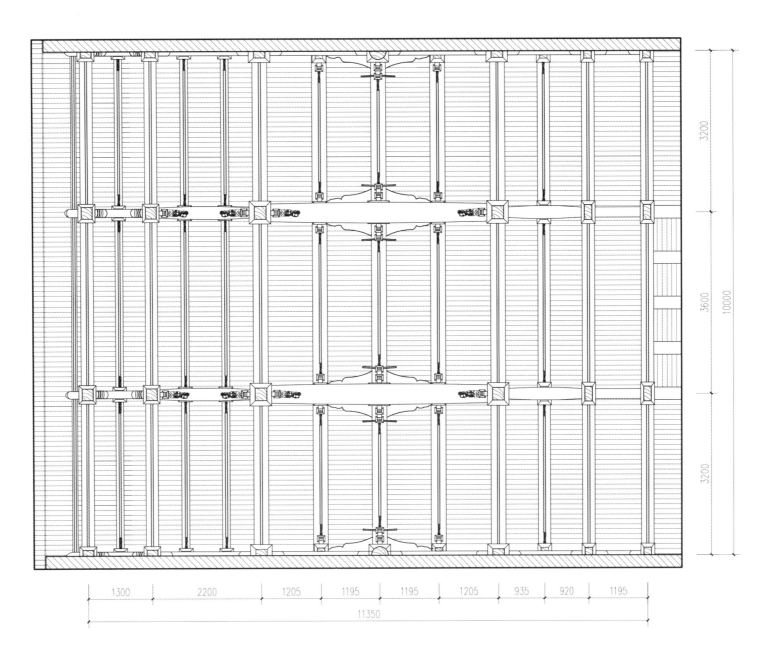

载酒堂梁架仰视图
Reflected truss plan: Zaijiu Tang

"诗酒联欢"墙门
Gate bearing the inscription 'Shijiu Lianhuan'

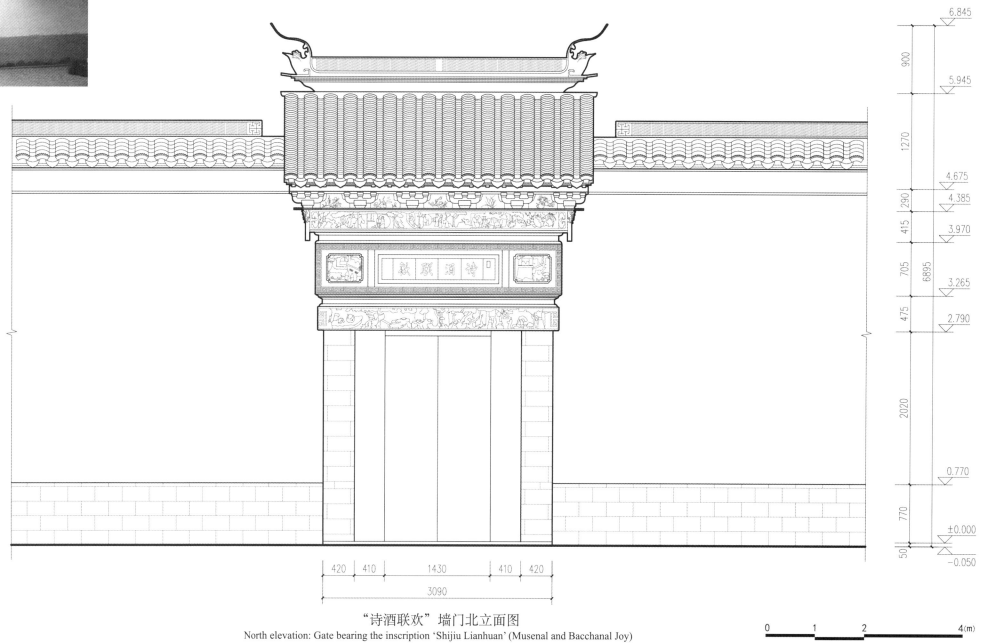

"诗酒联欢"墙门北立面图
North elevation: Gate bearing the inscription 'Shijiu Lianhuan' (Musenal and Bacchanal Joy)

1 无俗韵轩 Wu Suyun Xuan
2 城曲草堂 Chengqu Caotang
3 储香馆 Chuxiang Guan
4 藤花舫 Tenghua Fang
5 樨廊 Xi Lang
6 还砚斋 Huanyan Zhai
7 望月亭 Wangyue Ting
8 吾爱亭 Wu'ai Ting
9 山水间 Shanshui Jian
10 听橹楼 Tinglu Lou
11 魁星阁 Kuixing Ge

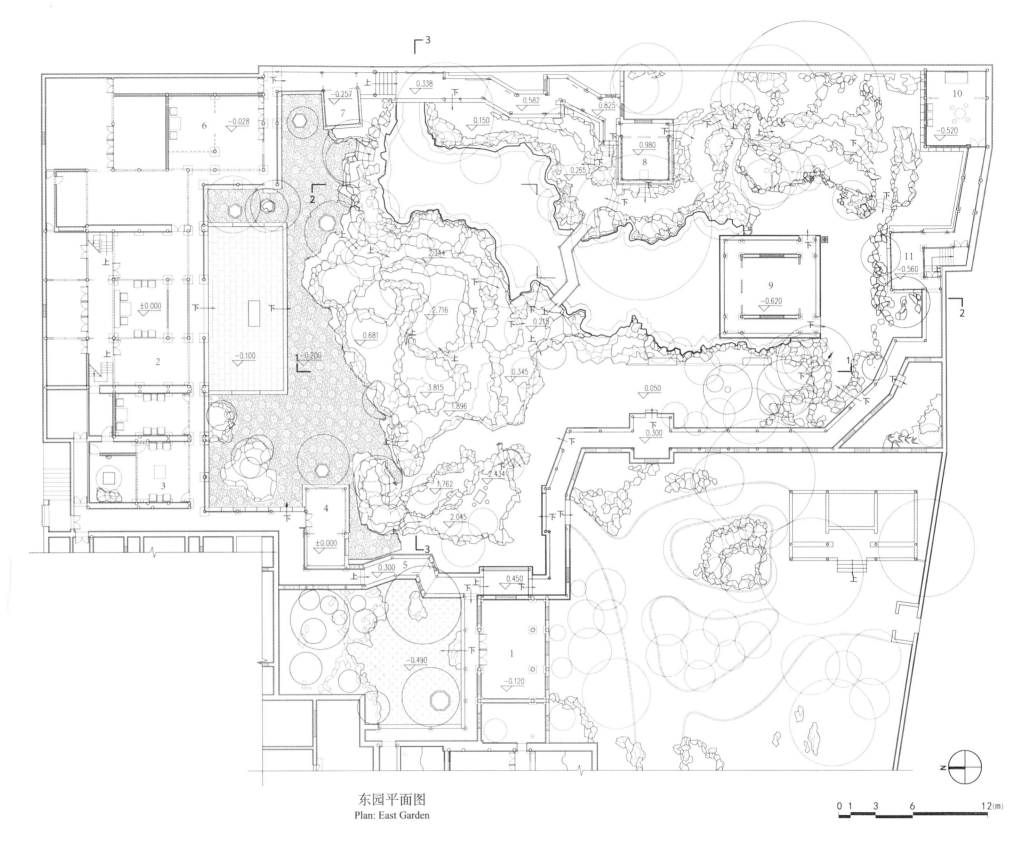

东园平面图
Plan: East Garden

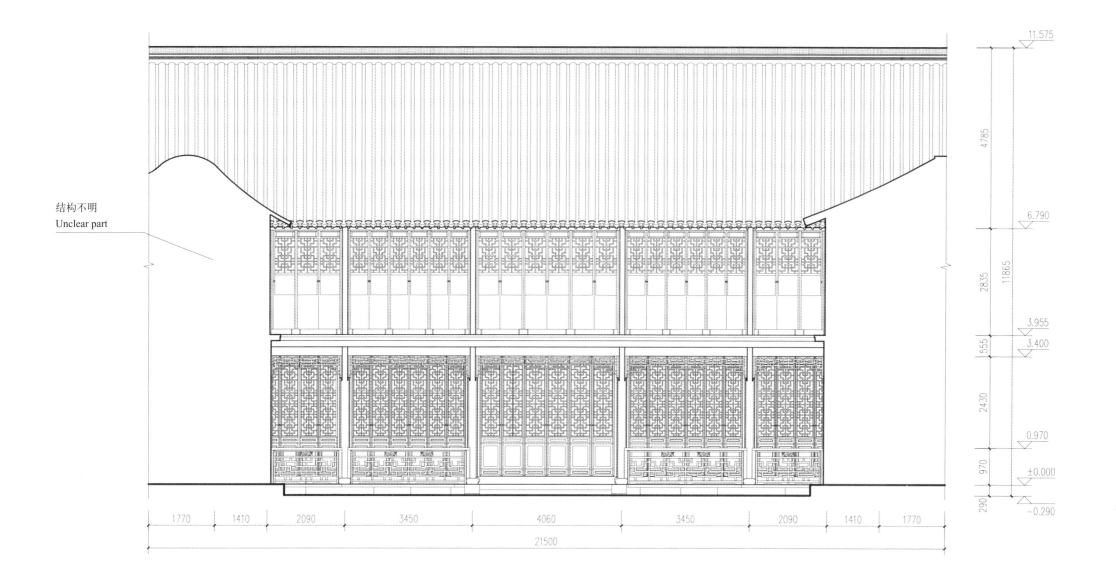

结构不明
Unclear part

楼厅南立面图
South elevation: Lou Ting

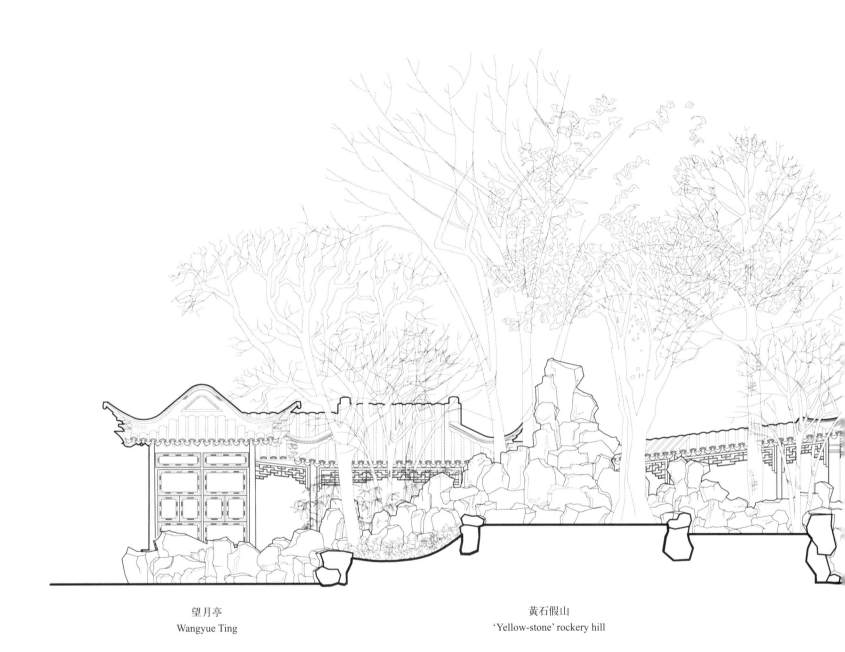

望月亭
Wangyue Ting

黄石假山
'Yellow-stone' rockery hill

东园 1-1 剖面图—假山南北剖东视
Section 1-1: East Garden / North-south section: Rockery hill

吾爱亭
Wu'ai Ting

山水间
Shanshui Jian

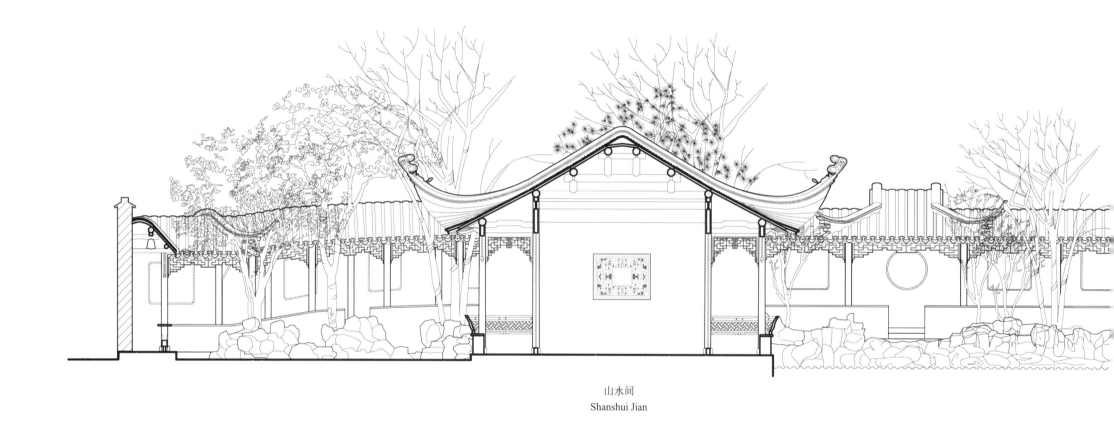

山水间
Shanshui Jian

东园 2-2 剖面图—假山南北剖西视
Section 2-2: East Garden / South-north section: Rockery hill

藤花舫
Tenghua Fang

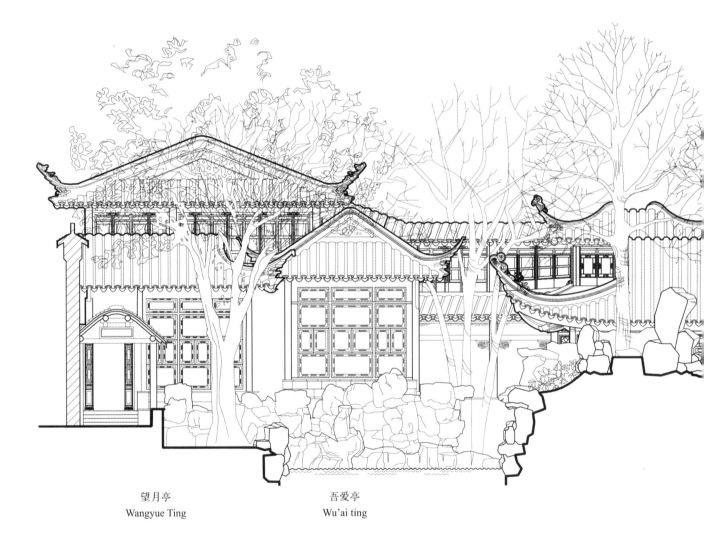

望月亭
Wangyue Ting

吾爱亭
Wu'ai ting

东园 3-3 剖面图—假山东西剖南视
Section 3-3: East Garden / East-west section: Rockery hill

黄石假山
'Yellow-stone' rockery hill

无俗韵轩院落
Wu Suyun Xuan: courtyard

无俗韵轩北面
Wu Suyun Xuan: North side

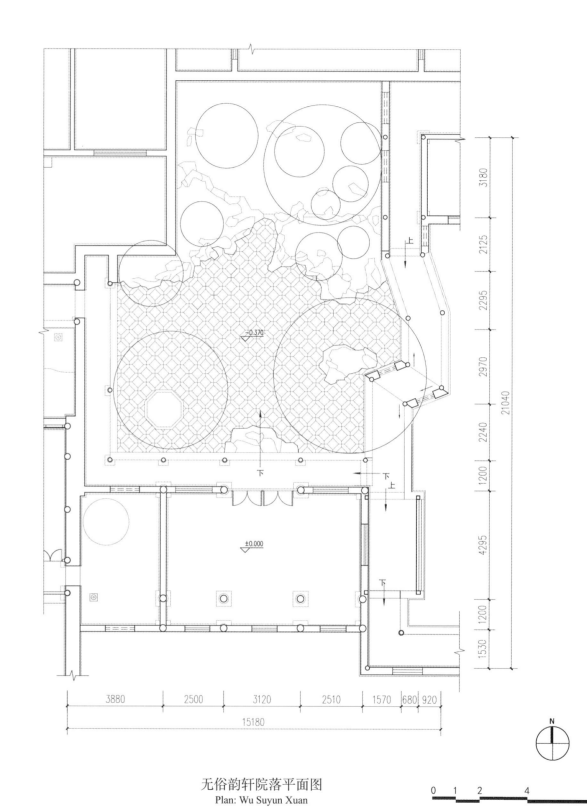

无俗韵轩院落平面图
Plan: Wu Suyun Xuan

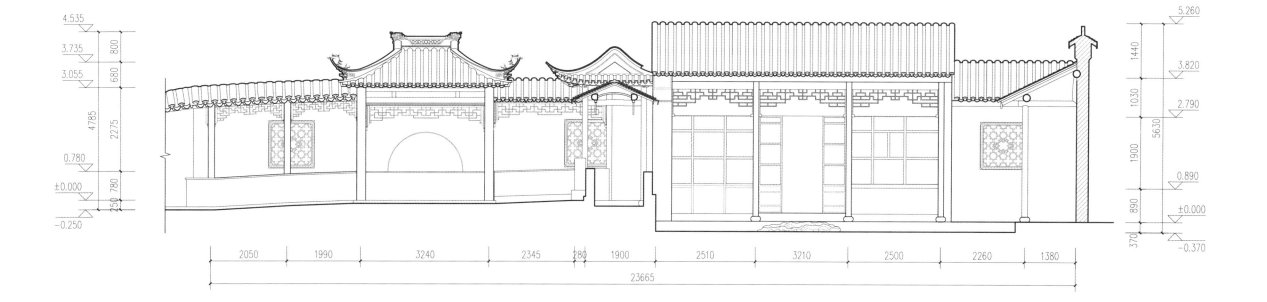

无俗韵轩院落北立面图
North elevation: Wu Suyun Xuan

城曲草堂南面
Chengqu Caotang: South side

城曲草堂室内
Chengqu Caotang: indoors

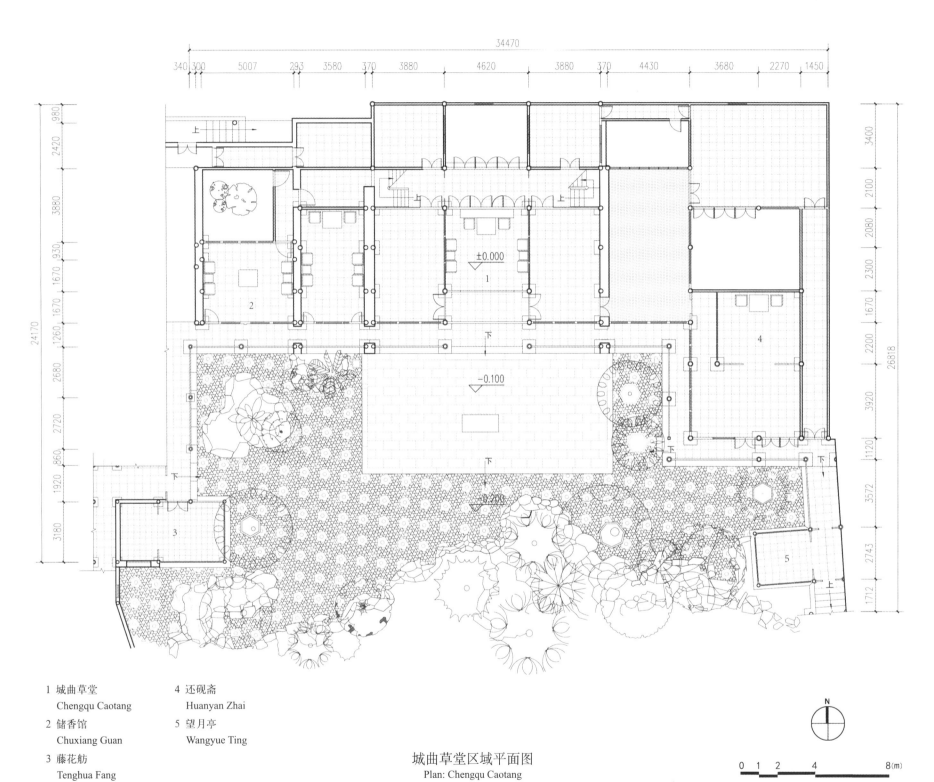

1 城曲草堂 Chengqu Caotang
2 储香馆 Chuxiang Guan
3 藤花舫 Tenghua Fang
4 还砚斋 Huanyan Zhai
5 望月亭 Wangyue Ting

城曲草堂区域平面图
Plan: Chengqu Caotang

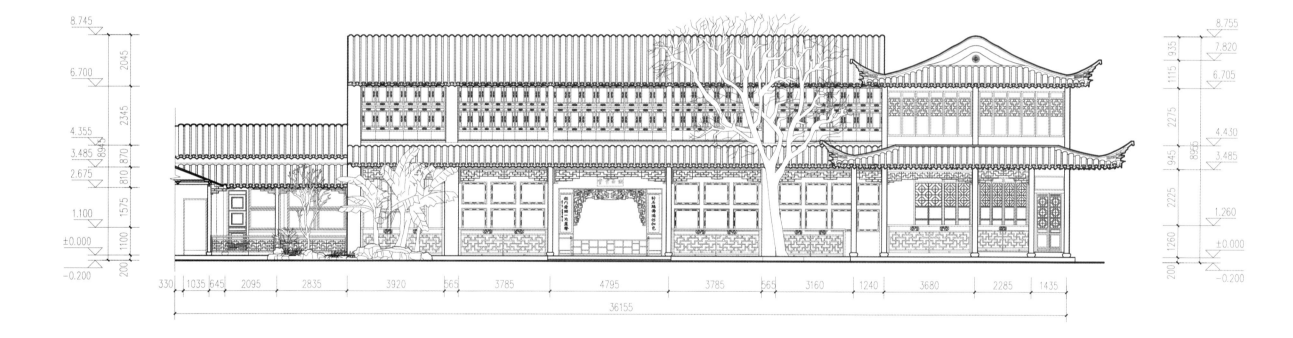

城曲草堂南立面图
South elevation: Chengqu Caotang

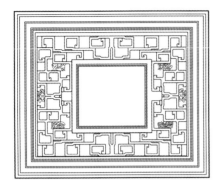

藤花舫立面窗大样图
Details: a window of Tenghua Fang

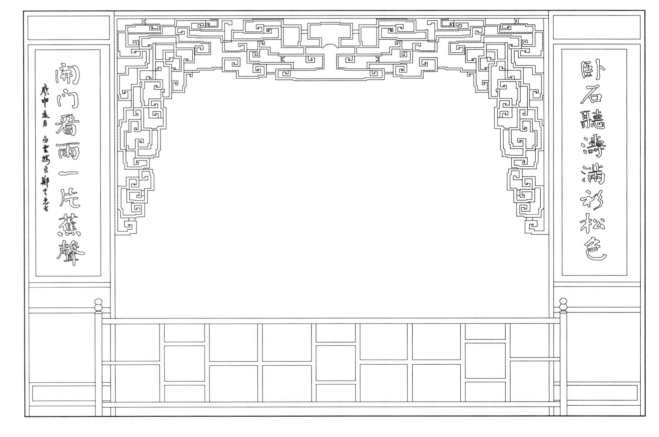

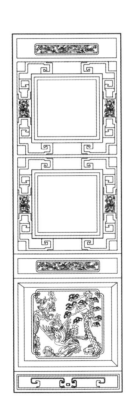

藤花舫格扇大样图
Details: a door of Tenghua Fang

储香馆格扇大样图
Details: a door of Chuxiang Guan

城曲草堂当心间立面格扇大样图
Details: *Gualuo* (hanging traceries), central-bay, Chengqu Caotang

城曲草堂区域模型
Model: Chengqu Caotang

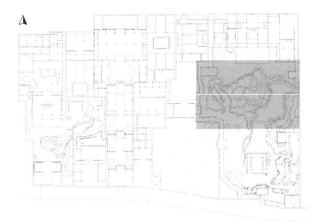

黄石假山
'Yellow-stone' rockery hill

1 藤花舫 Tenghua Fang
2 望月亭 Wangyue Ting

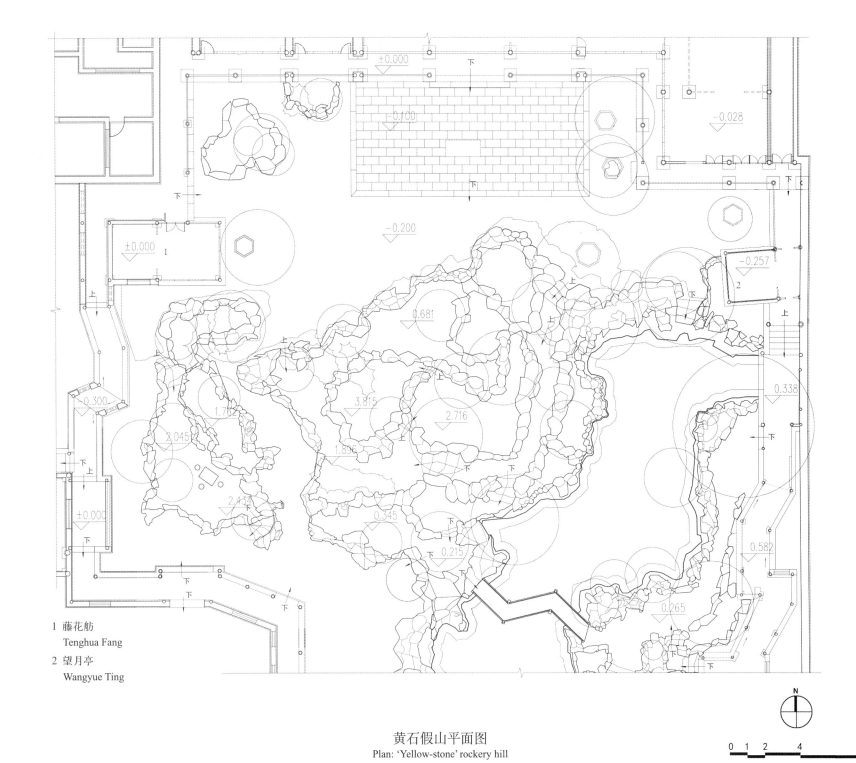

黄石假山平面图
Plan: 'Yellow-stone' rockery hill

| 望月亭 | 吾爱亭 | 山水间 | 黄石假山 |
| Wangyue Ting | Wu'ai Ting | Shanshui Jian | 'Yellow-stone' rockery hill |

黄石假山北立面图
North elevation: 'Yellow-stone' rockery hill

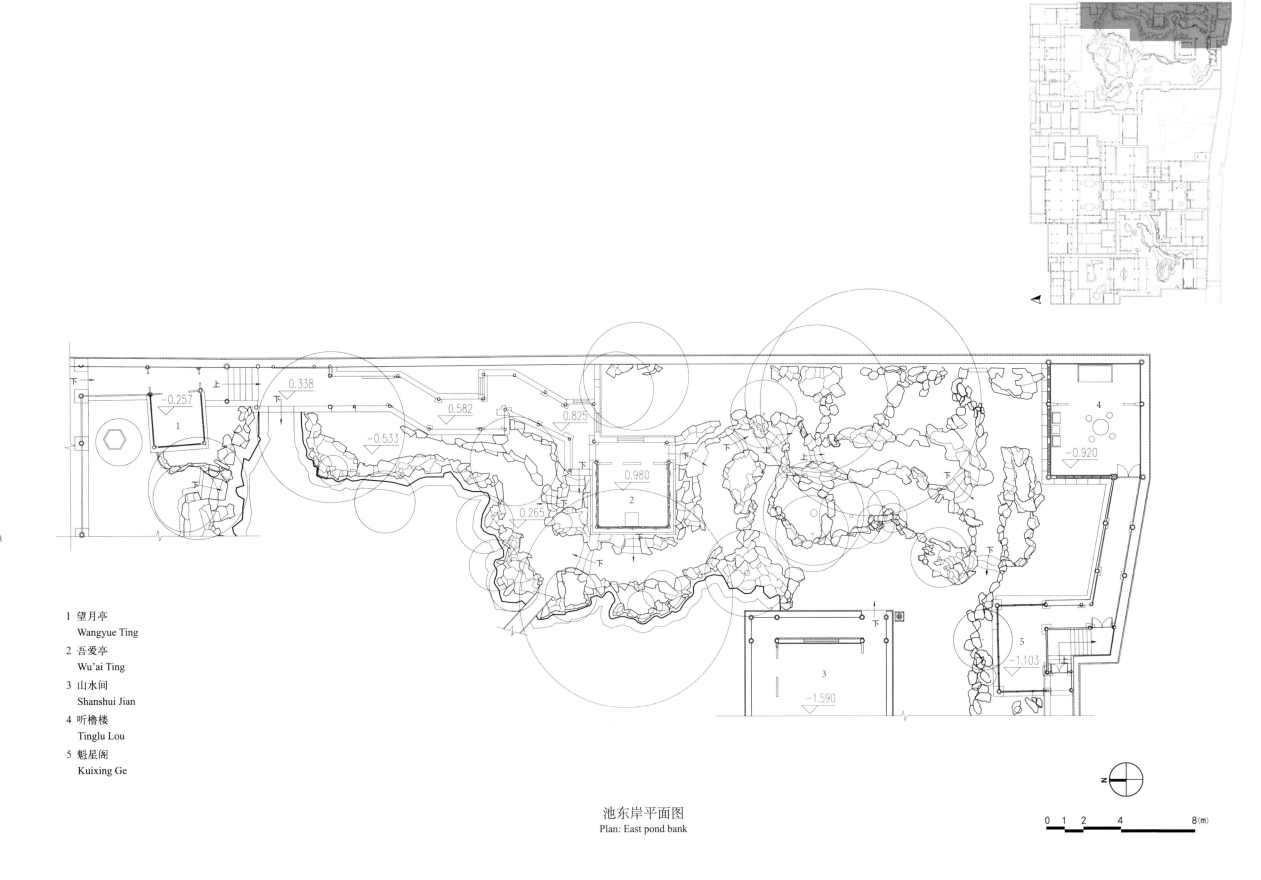

1 望月亭 Wangyue Ting
2 吾爱亭 Wu'ai Ting
3 山水间 Shanshui Jian
4 听橹楼 Tinglu Lou
5 魁星阁 Kuixing Ge

池东岸平面图
Plan: East pond bank

望月亭　Wangyue Ting　　　　吾爱亭　Wu'ai Ting

池东岸西立面图
West elevation: East pond bank

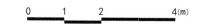

吾爱亭
Wu'ai Ting

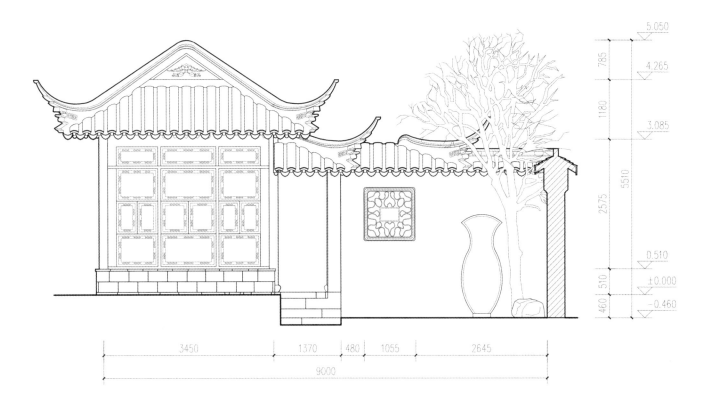

吾爱亭南立面图
South elevation: Wu'ai Ting

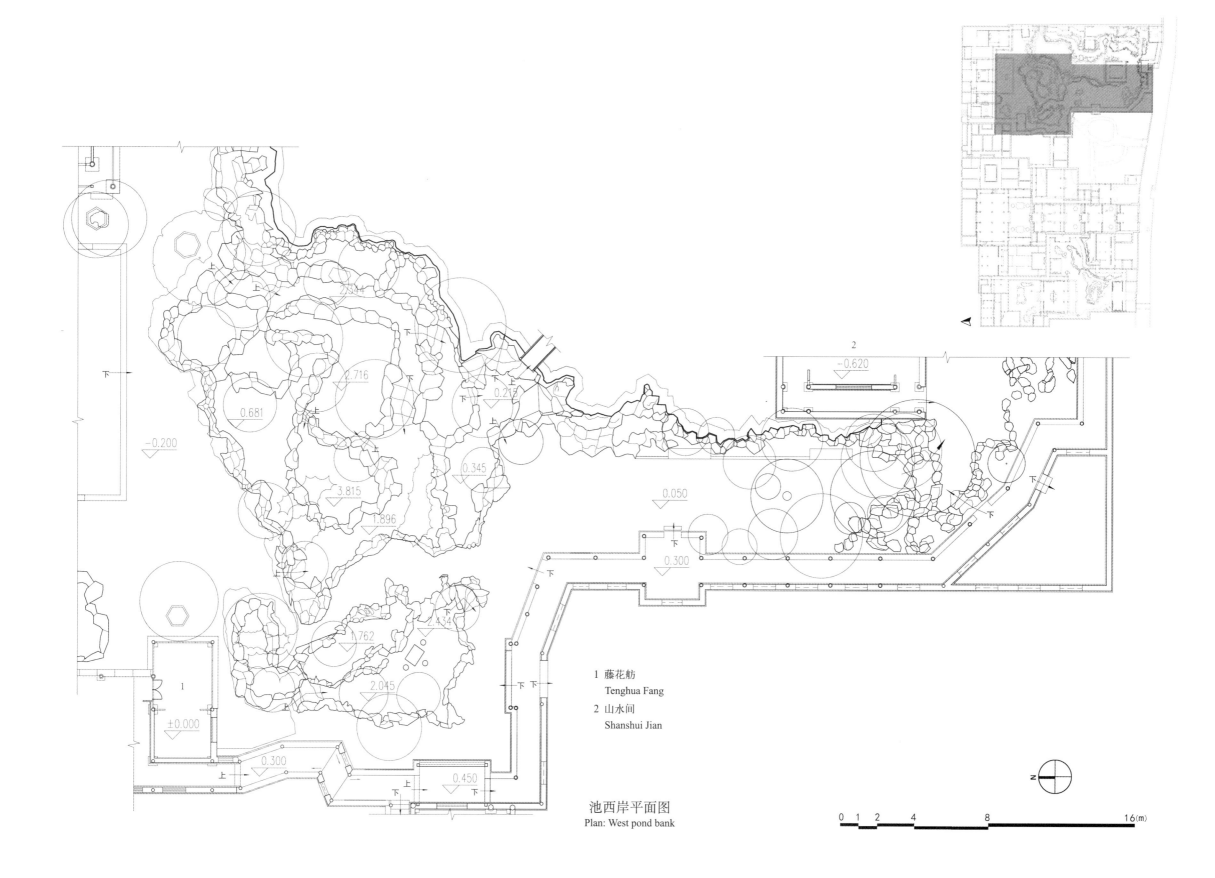

1 藤花舫
　Tenghua Fang
2 山水间
　Shanshui Jian

池西岸平面图
Plan: West pond bank

滴水大样图
Details: Flashing tiles

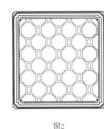
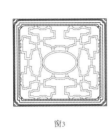

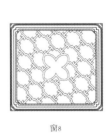

窗1　　窗2　　窗3　　窗4　　窗5　　窗6　　窗7　　窗8　　窗9　　窗10

廊子漏窗大样图
Details: *Louchuang* (traceried window-openings), galleries

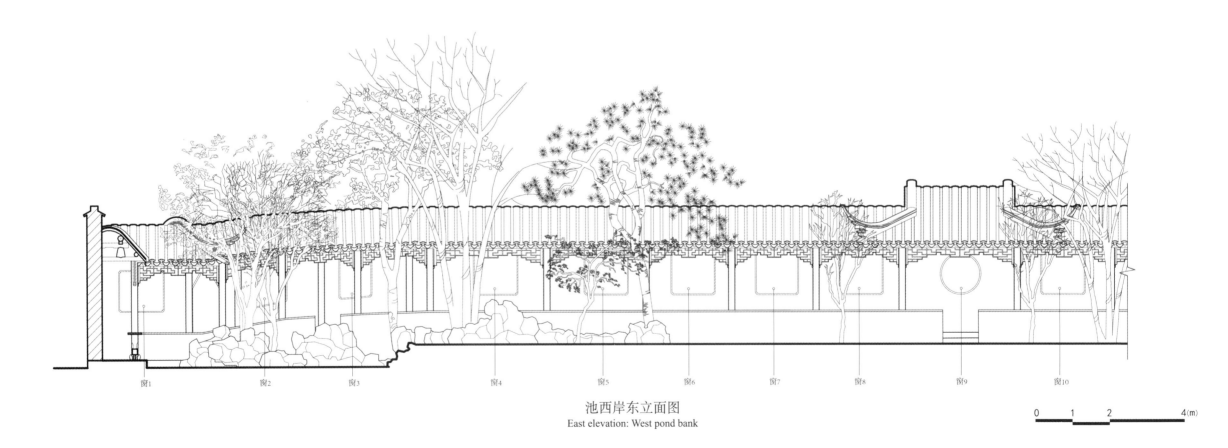

池西岸东立面图
East elevation: West pond bank

山水间
Shanshui Jian

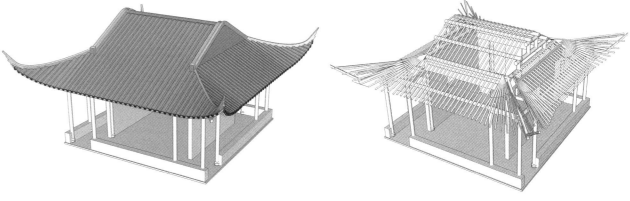

山水间模型
Model: Shanshui Jian

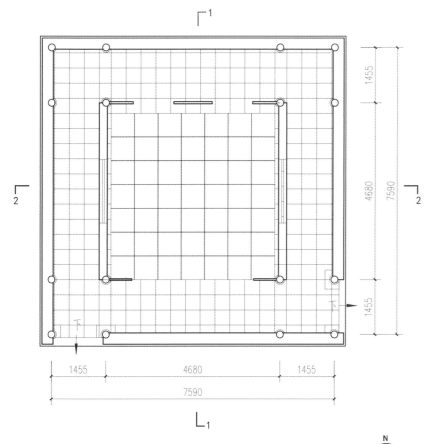

山水间平面图
Plan: Shanshui Jian

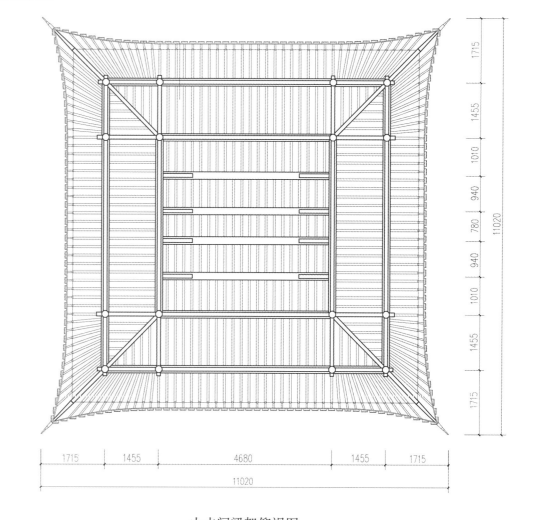

山水间梁架仰视图
Reflected truss plan: Shanshui Jian

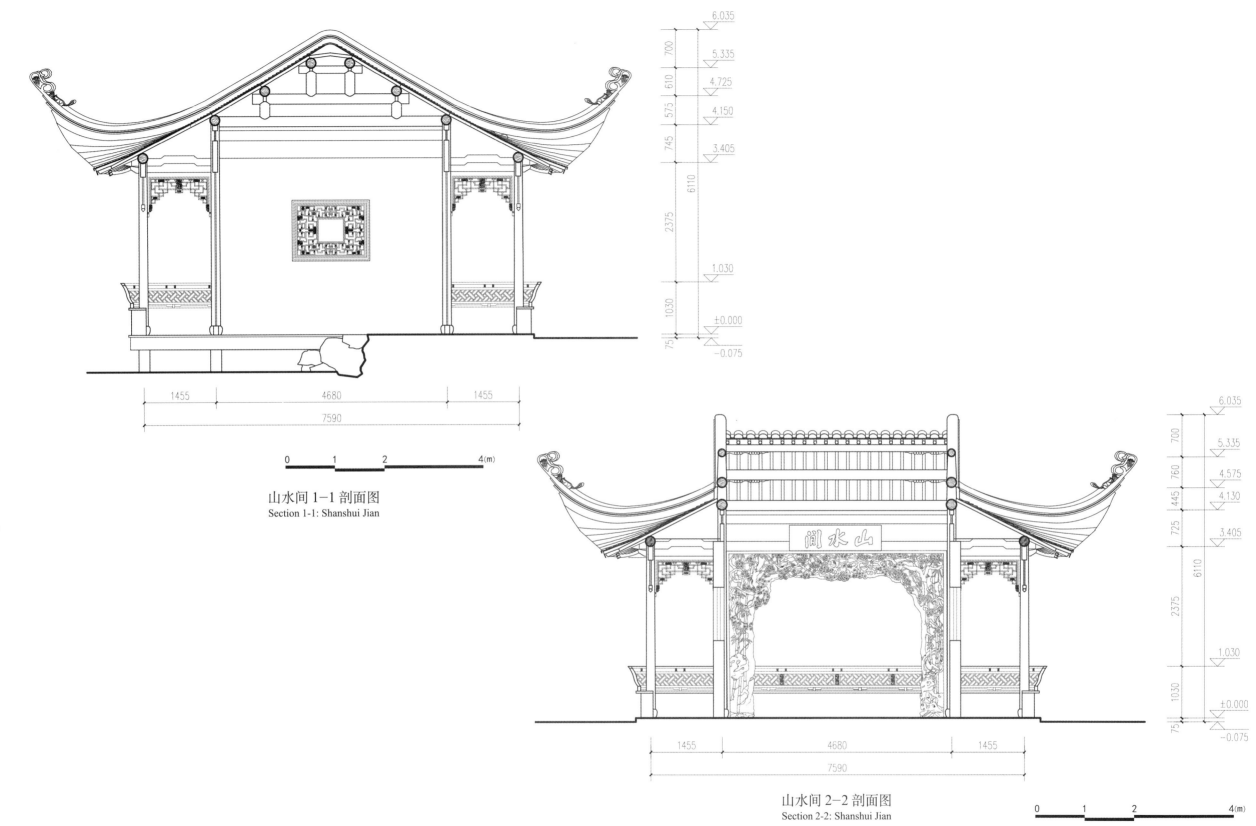

山水间 1-1 剖面图
Section 1-1: Shanshui Jian

山水间 2-2 剖面图
Section 2-2: Shanshui Jian

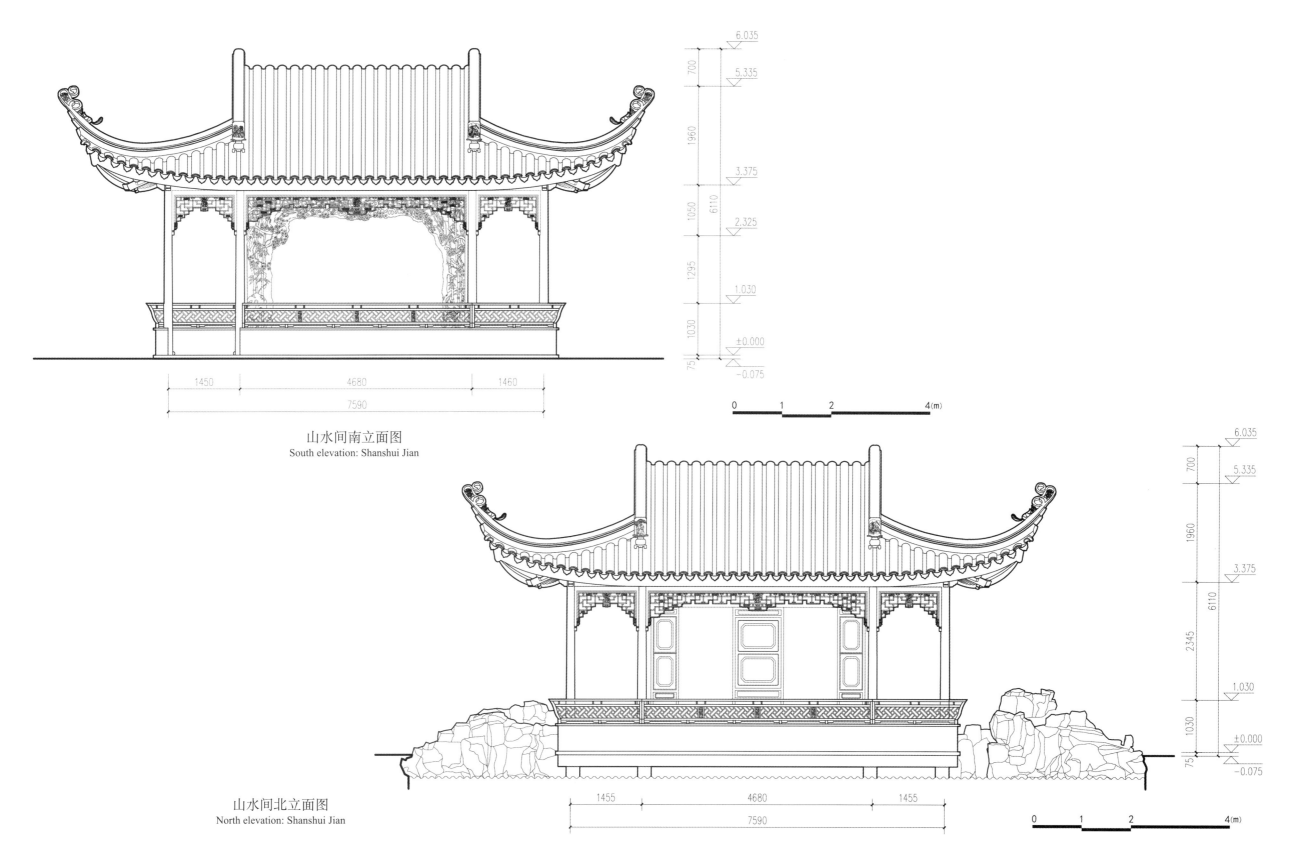

山水间南立面图
South elevation: Shanshui Jian

山水间北立面图
North elevation: Shanshui Jian

山水间格扇大样图
Details: *Geshan* (partition doors), Shanshui Jian

山水间落地罩大样图
Details: *Luodizhao* (traceried ceiling-to-floor screen), Shanshui Jian

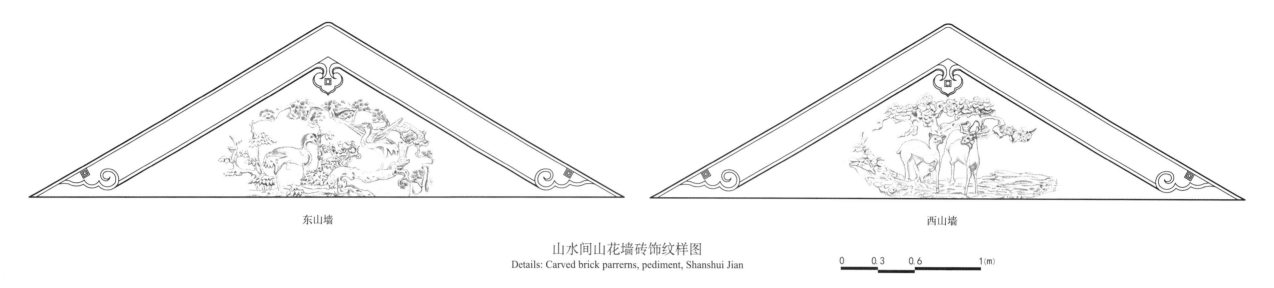

东山墙　　　　　　　　　　　　　　西山墙

山水间山花墙砖饰纹样图
Details: Carved brick parrerns, pediment, Shanshui Jian

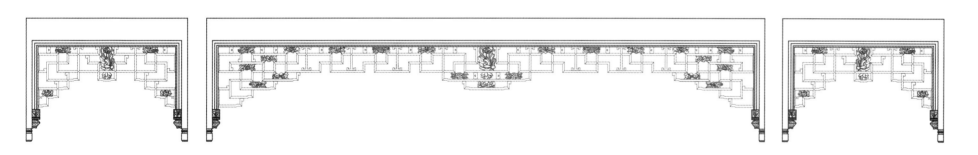

山水间挂落大样图
Details: *Gualuo* (traceried hanging fascia), Shanshui Jian

山水间山墙花窗大样图
Details: *Huachuang* (traceried window), gable wall, Shanshui Jian

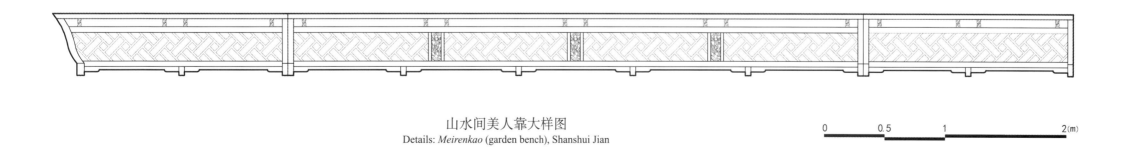

山水间美人靠大样图
Details: *Meirenkao* (garden bench), Shanshui Jian

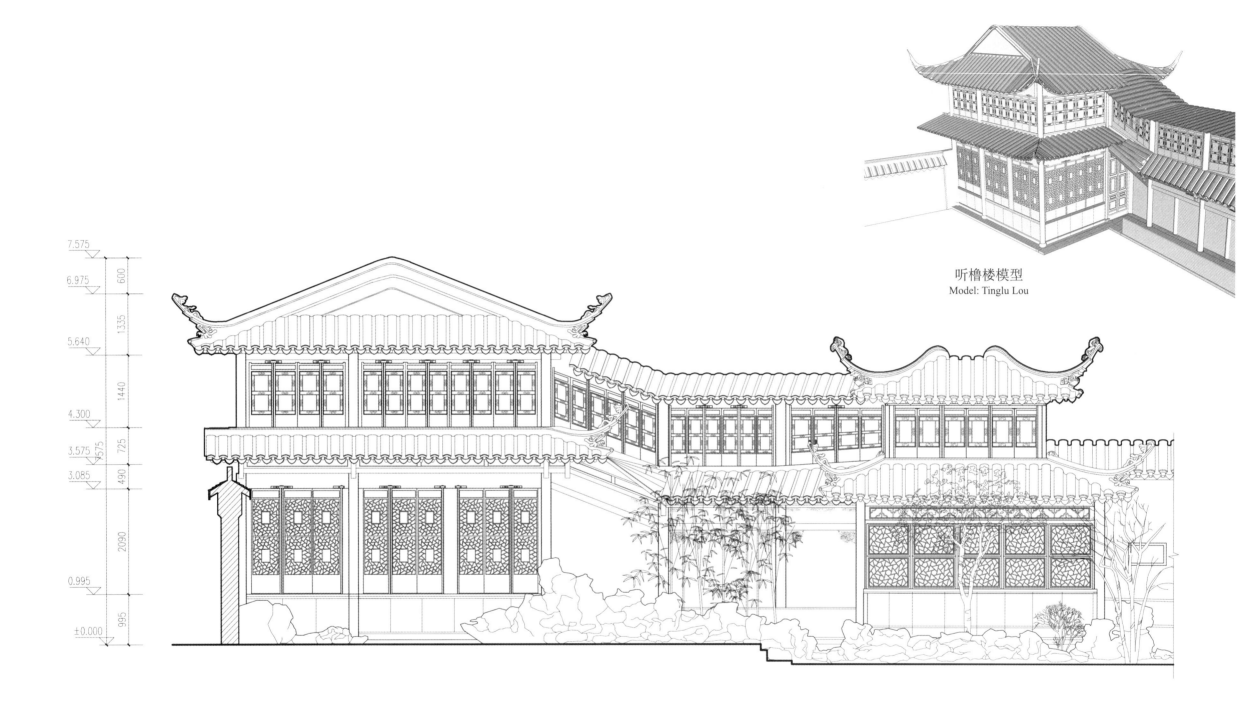

听橹楼模型
Model: Tinglu Lou

听橹楼－魁星阁北立面图
North elevation: Tinglu Lou and Kuixing Ge

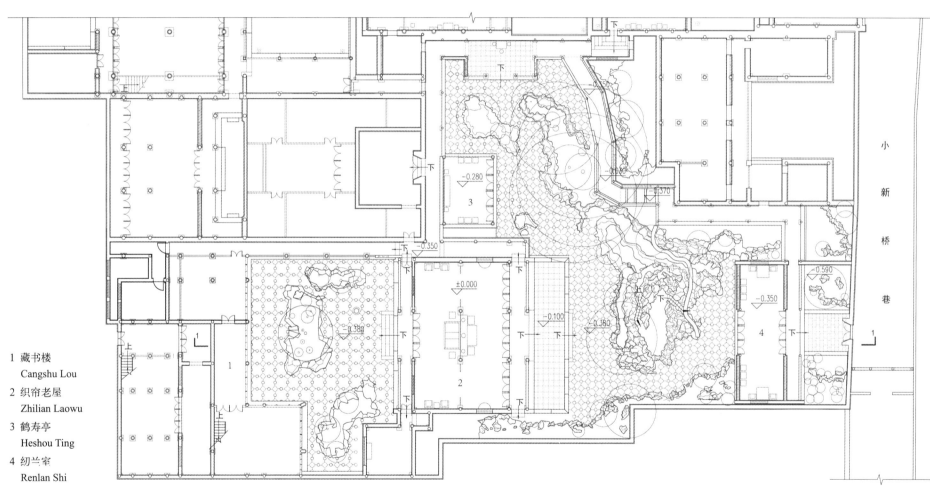

1 藏书楼 Cangshu Lou
2 织帘老屋 Zhilian Laowu
3 鹤寿亭 Heshou Ting
4 纫兰室 Renlan Shi

西园平面图
Plan: West Garden

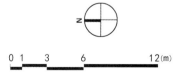

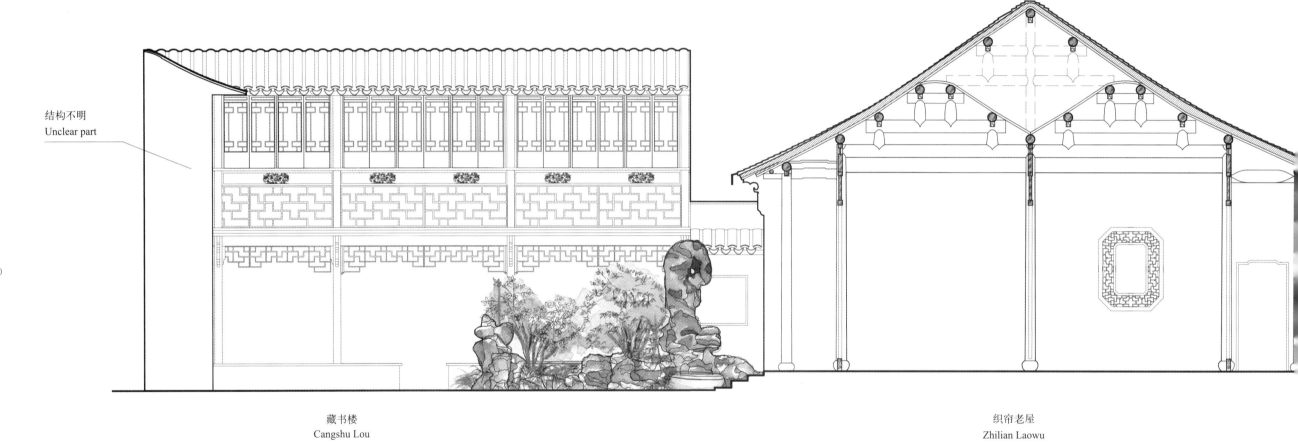

西园 1-1 剖面图
Section 1-1: West Garden

纫兰室院落
Renlan Shi: courtyard

织帘老屋
Zhilian Laowu

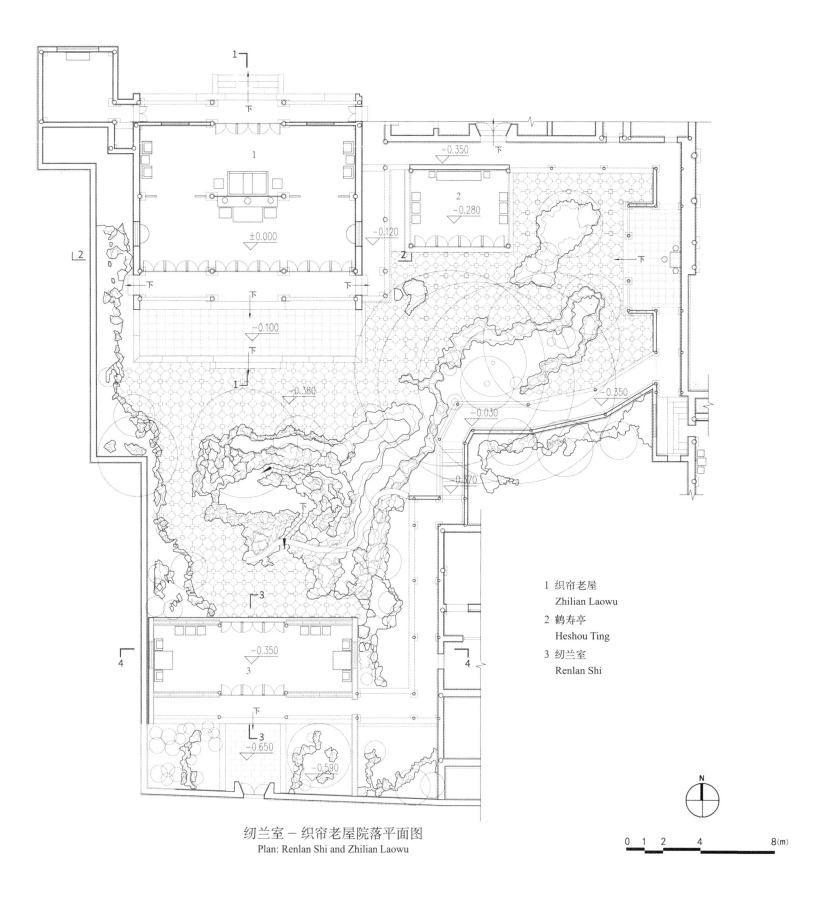

1 织帘老屋
  Zhilian Laowu

2 鹤寿亭
  Heshou Ting

3 纫兰室
  Renlan Shi

纫兰室－织帘老屋院落平面图
Plan: Renlan Shi and Zhilian Laowu

织帘老屋 1-1 剖面图
Section 1-1: Zhilian Laowu

织帘老屋模型
Model: Zhilian Laowu

织帘老屋梁架仰视图
Reflected Truss Plan: Zhilian Laowu

织帘老屋 2-2 剖面图
Section 2-2: Zhilian Laowu

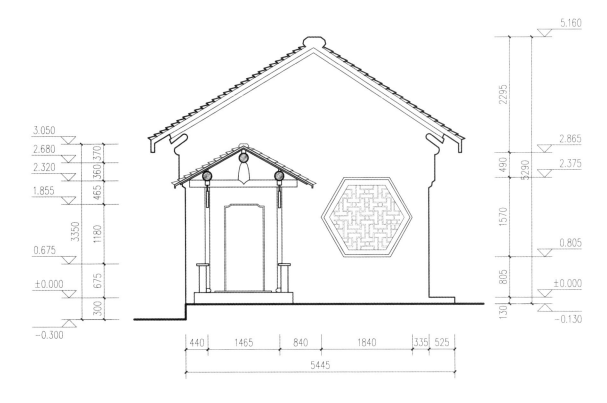

纫兰室东立面图
East elevation: Renlan Shi

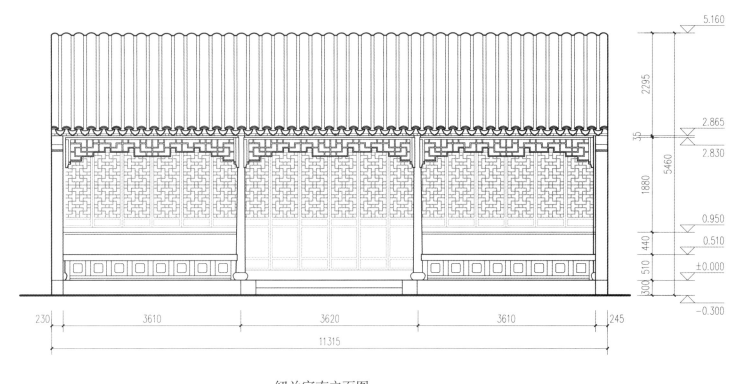

纫兰室南立面图
South elevation: Renlan Shi

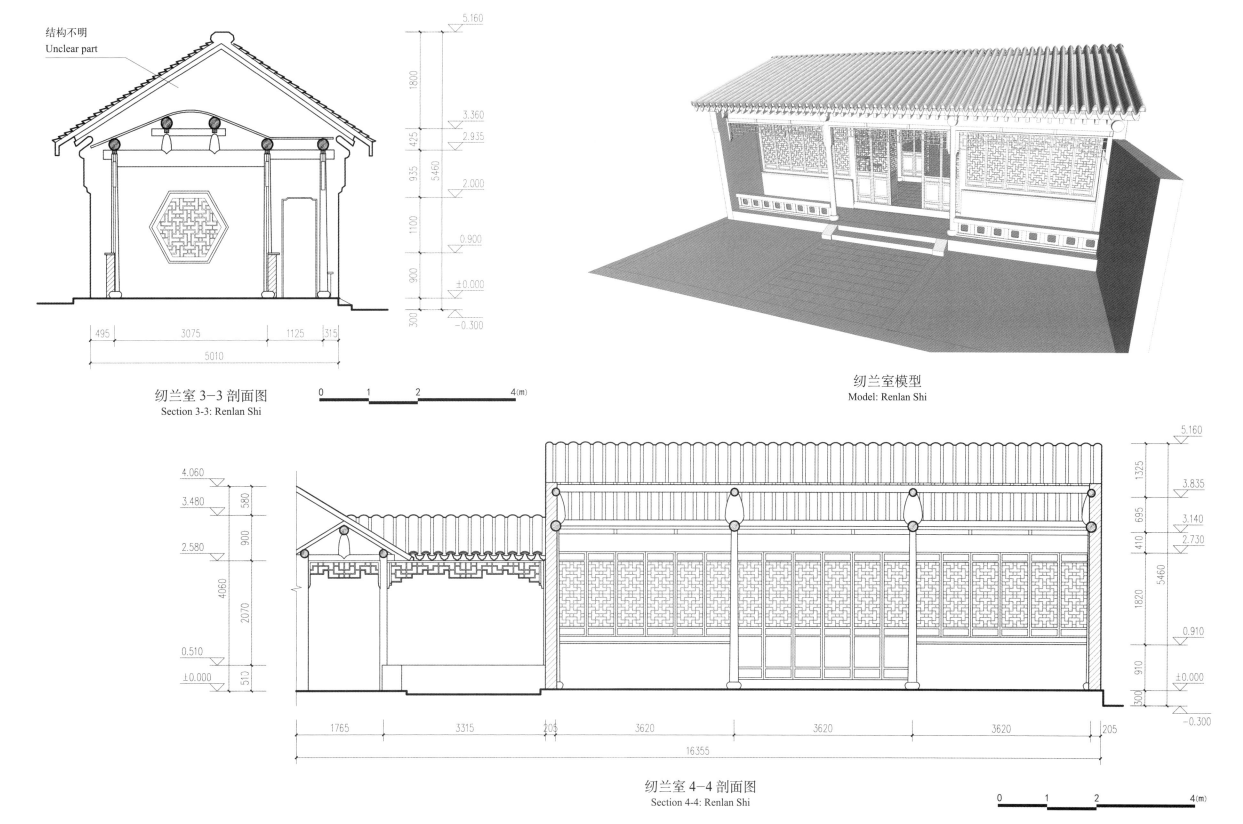

纫兰室 3-3 剖面图
Section 3-3: Renlan Shi

纫兰室模型
Model: Renlan Shi

纫兰室 4-4 剖面图
Section 4-4: Renlan Shi

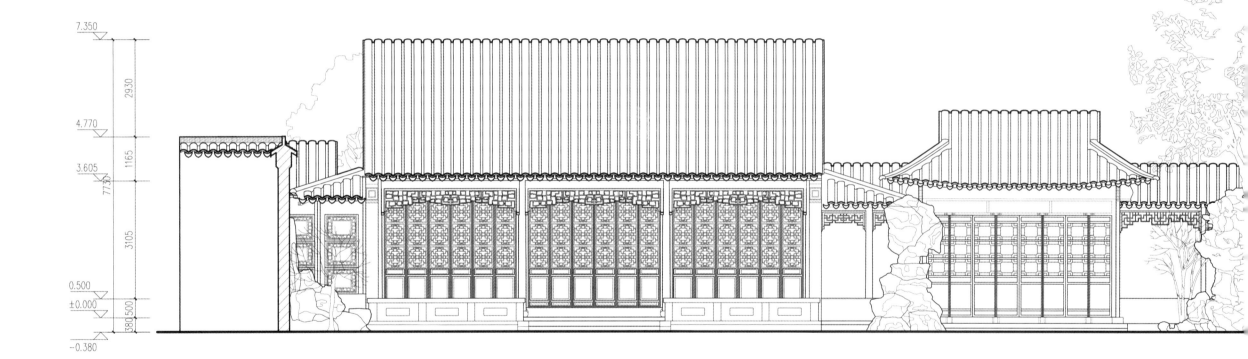

织帘老屋－鹤寿亭南立面图
South elevation: Zhilian Laowu and Heshou Ting

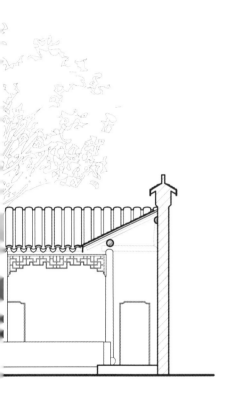
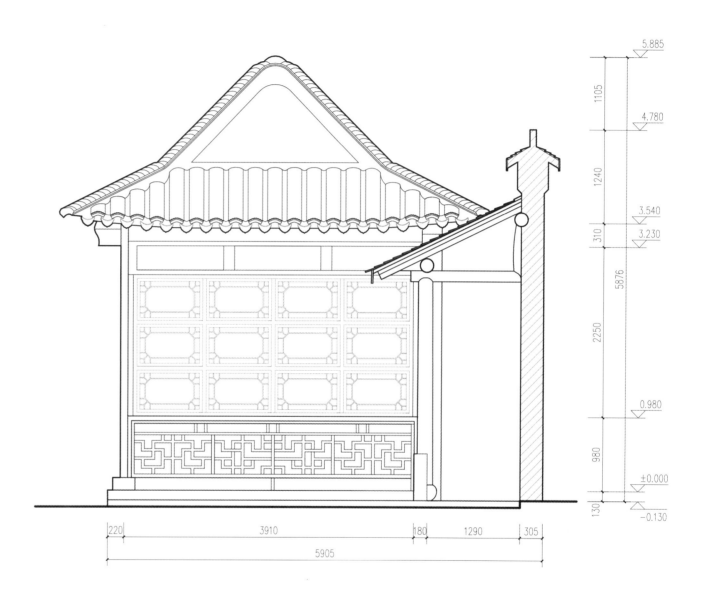

鹤寿亭东立面图
East elevation: Heshou Ting

西园假山南立面图
South elevation: West Garden rockery hill

西园假山北立面图
North elevation: West Garden rockery hill

藏书楼院落
Cangshu Lou: courtyard

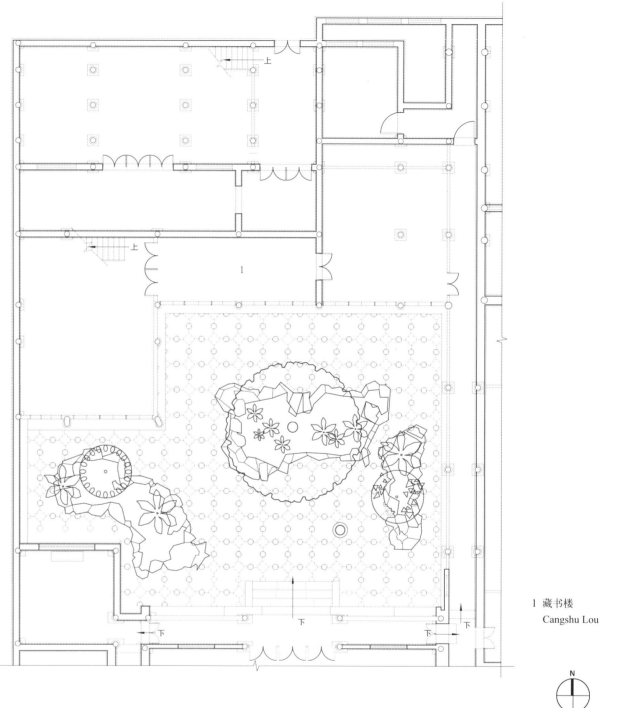

1 藏书楼
　Cangshu Lou

藏书楼院落平面图
Plan: Cangshu Lou

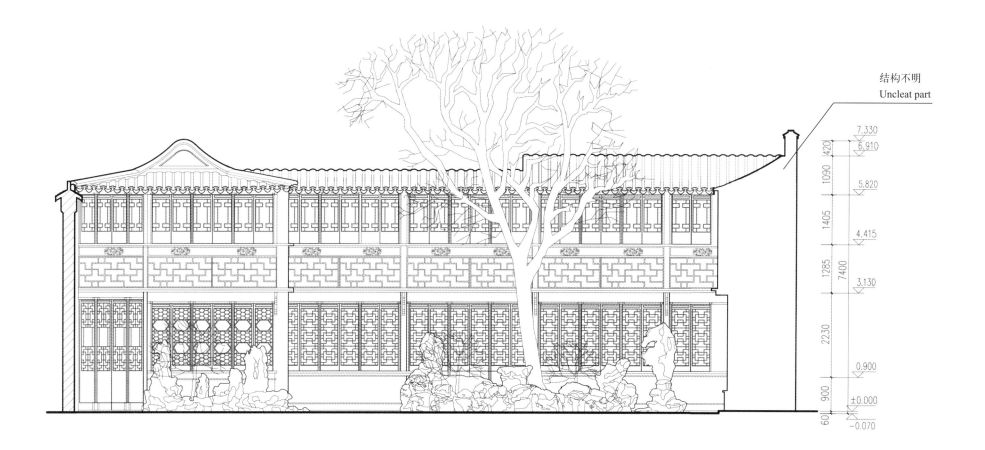

藏书楼南立面图
South elevation: Cangshu Lou

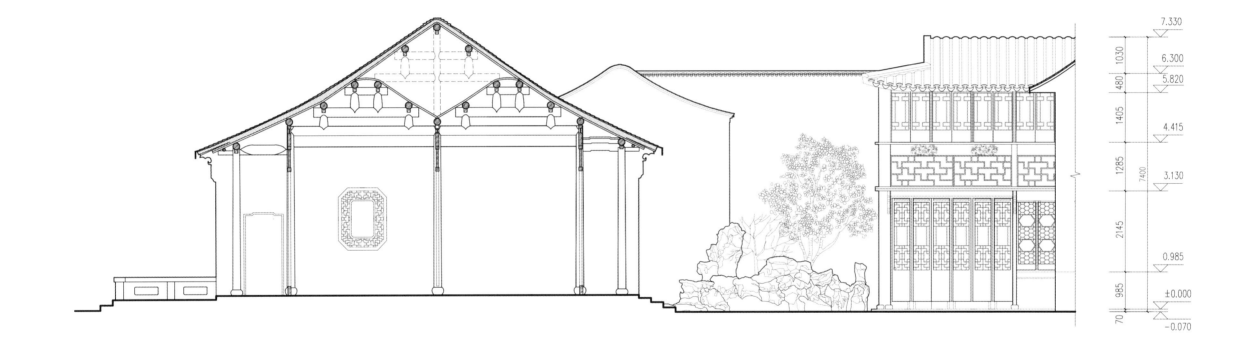

织帘老屋 3-3 剖面—藏书楼东立面图
Section 3-3: Zhilian Laowu / East elevation: Cangshu Lou

苏州怡园
Yi Yuan (Garden of Delight), Suzhou

· 地址：苏州市人民路中段原 43 号（现 1265 号）
· 年代：清
· 保护级别：江苏省文物保护单位

· Address: 1265 Renmin Rd., Gusu Dist., Suzhou
· Origin: 19th Century
· Status: Cultural Heritage Site under Provincial Protection (1982)

04 苏州怡园

苏州怡园位于苏州老城居中偏西的乐桥路上，东为护龙街（现人民路），西与原顾氏春荫义庄及祠堂毗邻，南与原顾氏住宅隔巷相对，北为子弹巷。清同治末年至光绪初年（1874—1882年），怡园由浙江宁绍道台顾文彬及其子顾承主持修造。园中盛时曾举行画社、琴会、曲会等雅集活动，后战乱频仍，园中建筑、陈设因而损毁。1953年，怡园由顾氏后人连同家祠一并捐为国有，并经修复后于当年12月正式开放以供游赏。入园本须经由春荫义庄，修复后则于人民路上新设园门。怡园于1963年列入第一批苏州市文物保护单位，后又于1982年提升为江苏省文物保护单位。

全园占地面积现约九亩，分为东西两部分，之间以复廊相隔。东部以建筑为主，主要包括玉延亭、四时潇洒亭、长廊围合而成的院落和拜石轩、坡仙琴馆院落；西部则为池山区，水池居中，池北筑山，池南藕香榭为园内主厅，榭北设临池平台。榭东接长廊和南雪亭，榭西有廊与碧梧栖凤馆相通；池西部水面狭长，西岸为画舫斋，斋西为湛露堂。主山分为东西两部分，之间以小径相隔。山西南最高处建有螺髻亭，东北亦有亭，名"小沧浪"。

图1：入口部分
Fig.1: Entrance area

图2：玉延亭－四时潇洒亭
Fig.2: Yuyan Ting and Sishi Xiaosa Ting

主要参考文献：
[1] 苏州市园林和绿化管理局.怡园志[M].文汇出版社，2013.

测　　绘：
西安建筑科技大学建筑学院建筑学2010级02，04班——
张博、何薇、何笑、洪毅、刘兴东、徐尚哲；徐丹、李唱、车林津、吴昊、吕抱朴、薛超；王新蕊、焦子倩、徐沛豪、李天宇、刘征、周振宇、徐冰；蔡瑾、李宇轩、卢凯、闫丹妮、赵文博、赵英男、鞠曦、虞聪；崇显鹏、冯越、高泽华、张晓兰、张学毓、张也、张瀛、仪若瑜、闫睿婧；徐奥文、张浩、张强、姚瑶、周梁少强、杨东朴；苗常茂、徐晓捷、张琼、朱一鹤、徐原野；吴明奇、李野墨、王诗宇、赵若菡、周小丁、王经纬、陈楚康、段言泽、葛中斌、路嘉君、朱苾均；戴茜、冯贞珍、任飞、唐宇峰、王欣兰；程华旸、董卓越、王嘉炜、张汀兰、李泱

指导教师：喻梦哲、张文波；
图纸绘制、整理：王茹悦、魏顾；
测绘时间：2013年10月

04. Yi Yuan (Garden of Delight)

Yi Yuan sits on Leqiao Rd., in mid-west historic urban area of Suzhou. It is embraced by historically-rich fabrics including Hulong Jie (Dragon-Guarding Street) (present-day Renmin Road), on its east, Danzi Xiang (Slingshot Pellet Lane), north, former Gu family's Chunyin Yizhuang (Spring-Shade Charity House) and ancestral shrine, west, along with Gu's former residence across the lane south. Between 1874 and 1882, Gu Wenbin (1811–1889), a former high-ranking official of Zhejiang Province, and his son Cheng commissioned the construction of the Garden. Back in its prime, the site served as a salon for painters, musicians, and dramatists until damages were inflicted upon its structures and amenities by the warfare, followed by decline and dilapidation. In 1953 Gu's descendants transferred, by donation, the site to a state-owned property. This status change led to an immediate government-commissioned project of restoration, of which upon the completion the Garden was reopened to the public in December the very same year. Also by this project a new entrance from Renmin Rd. was created, as opposed to the original one shared by Chunyin Yizhuang. In 1963 the Garden joined the ranks of the earliest Cultural Heritage Sites under Municipal Protection and in 1982 was upgraded to provincial level.

Now the nearly 6,000m^2 garden could be divided into two parts, separated from each other by a set of *fulang* (both-sides-opening galleries). The eastern part is composed mainly by garden structures, including Yuyan Ting (Yam Pavilion), Sishi Xiaosa Ting (Eternal-Elegance Pavilion), a courtyard enclosed by galleries, and another courtyard featuring Baishi Xuan (Stone-Saluting Pavilion) and Poxian Qinguan (Po-Xian[1] Musical House). The western part is largely occupied by rockworks and a pond. The pond is placed at the center, of which rockworks sit on the north bank while Ou-Xiang Xie (Fragrant Lotus-Roots Hall), the main hall of the garden, stands on the south bank. The hall has a waterside platform protruding northward out over the pond. It is also connected by galleries on both sides, east to Nanxue Ting (Plum-Blossom Pavilion) and west to Biwu Qifeng Guan (Phoenix-Perch House). The west part of the pond has a long and narrow layout, of which on the west bank stands Huafang Zhai (Pleasure-Boat Study) and Zhanlu Tang (Dense Dewdrop Hall). The rockery hill is also divided longitudinally in half by sinuous paths, on the highest point of which at its southwest part stands Luoji Ting (Snail-Shaped Topknot Pavilion). The northeast part has another pavilion named Xiao Canglang (Minor Canglang Pavilion).

图 3：藕香榭
Fig.3: Ouxiang Xie

图 4：面壁亭
Fig.4: Mianbi Ting

图 5：螺髻亭
Fig.5: Luoji Ting

Notes:

1. Po-Xian refers to SU Shi (1037-1101), a.k.a. SU Dongpo, a famous writer, poet, artist, calligrapher, scholar and politician of North Song Dynasty (960-1127), who has been treated as an apotheotic cultural figure in historical Chinese narratives.

Bibliography

Suzhou Municipal Bureau for Gardens and Forestation. (2013) *Yi Yuan Zhi* (Records of the Garden of Delight). Shanghai: Wenhui Press (in simpliefied Chinese).

Credits

Surveyors:

2015 Undergraduate Class B & D of Architecture (in groups):

ZHANG Bo, HE Wei, HE Xiao, HONG Yi, LIU Xingdong, XU Shangzhe; XU Dan, LI Chang, CHE Linjin, WU Hao, LV Baopu, XUE Chao; WANG Xinrui, JIAO Ziqian, XU Peihao, LI Tianyu, LIU Zheng, ZHOU Zhenyu, XU Bing; CAI Jin, LI Yuxuan, LU Kai, YAN Danni, ZHAO Wenbo, ZHAO Yingnan, JU Xi, YU Cong; CHONG Xianpeng, FENG Yue, GAO Zehua, ZHANG Xiaolan, ZHANG Xueyu, ZHANG Ye, ZHANG Ying, YI Ruoyu, YAN Ruijing; XU Aowen, ZHANG Hao, ZHANG Qiang, YAO Yao, ZHOU Liangshaoqiang, YANG Dongpu; MIAO Changmao, XU Xiaojie, ZHANG Qiong, ZHU Yihe, XU Yuanye; WU Mingqi, LI Yemo, WANG Shiyu, ZHAO Ruohan, ZHOU Xiaoding, WANG Jingwei; CHEN Chukang, DUAN Yanze, GE Zhongbin, LU Jiajun, ZHU Lijun; DAI Qian, FENG Zhenzhen, REN Fei, TANG Yufeng, WANG Xinlan; CHENG Huayang, DONG Zhuoyue, WANG Jiawei, ZHANG Tinglan, LI Yang.

Tutors: YU Mengzhe, ZHANG Wenbo

Drawing Preparation: WANG Ruyue, WEI Qi

Date: October, 2013

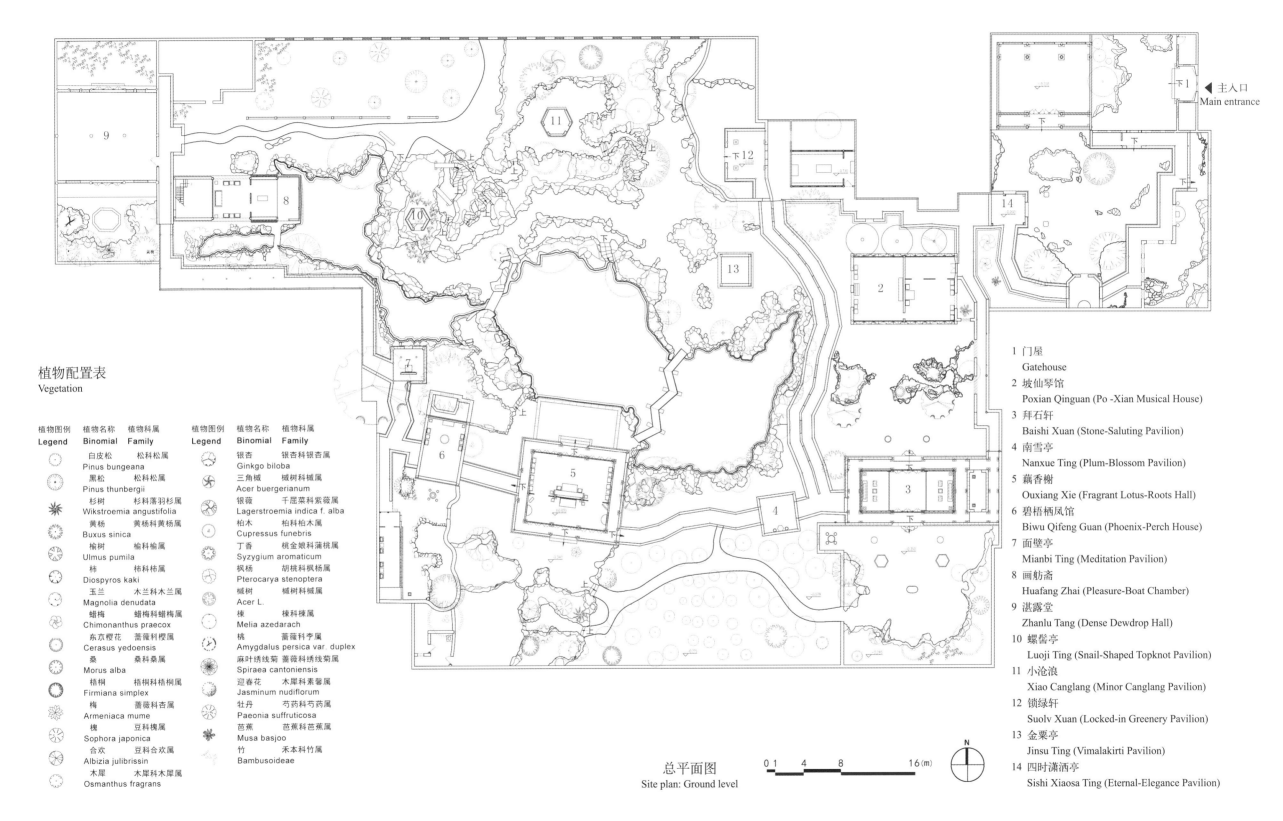

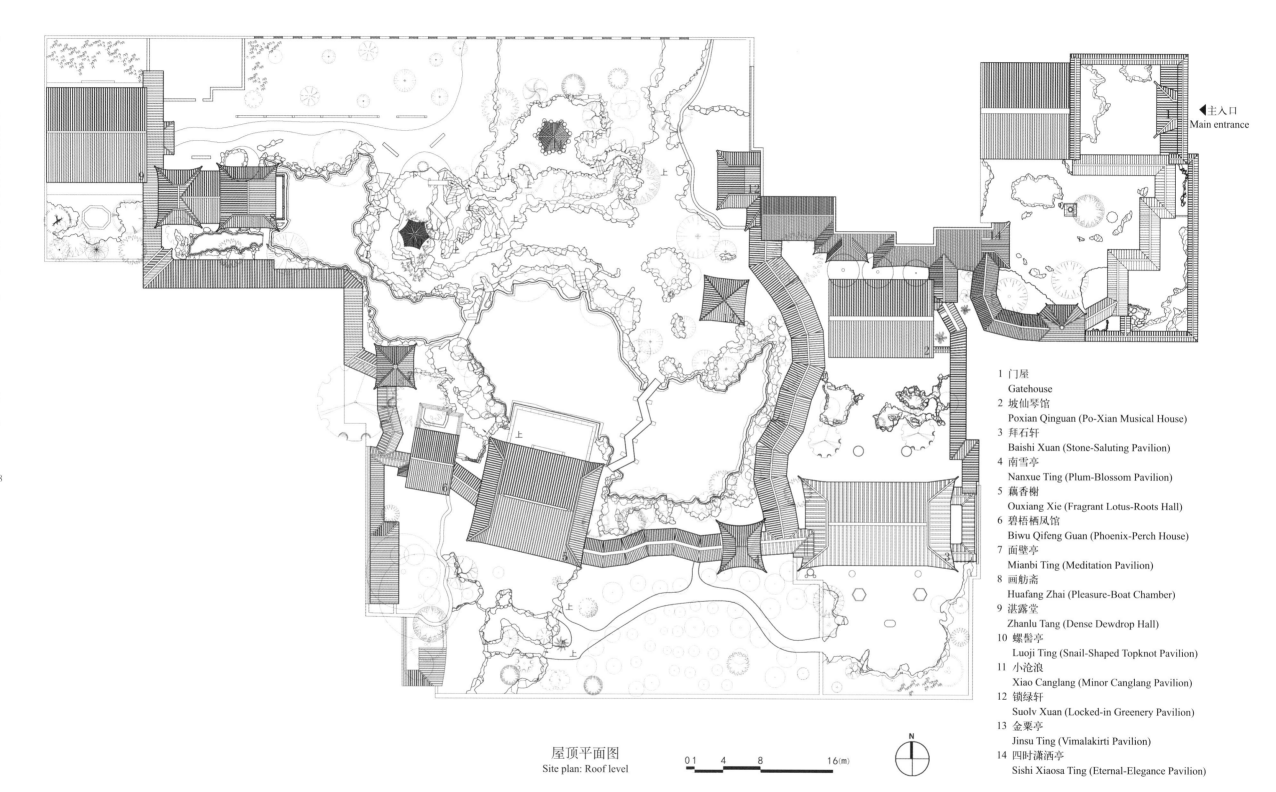

屋顶平面图
Site plan: Roof level

Main entrance 主入口

1 门屋 Gatehouse
2 坡仙琴馆 Poxian Qinguan (Po-Xian Musical House)
3 拜石轩 Baishi Xuan (Stone-Saluting Pavilion)
4 南雪亭 Nanxue Ting (Plum-Blossom Pavilion)
5 藕香榭 Ouxiang Xie (Fragrant Lotus-Roots Hall)
6 碧梧栖凤馆 Biwu Qifeng Guan (Phoenix-Perch House)
7 面壁亭 Mianbi Ting (Meditation Pavilion)
8 画舫斋 Huafang Zhai (Pleasure-Boat Chamber)
9 湛露堂 Zhanlu Tang (Dense Dewdrop Hall)
10 螺髻亭 Luoji Ting (Snail-Shaped Topknot Pavilion)
11 小沧浪 Xiao Canglang (Minor Canglang Pavilion)
12 锁绿轩 Suolv Xuan (Locked-in Greenery Pavilion)
13 金粟亭 Jinsu Ting (Vimalakirti Pavilion)
14 四时潇洒亭 Sishi Xiaosa Ting (Eternal-Elegance Pavilion)

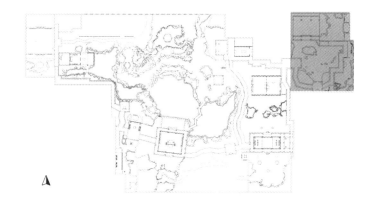

入口院落
Entrance courtyard

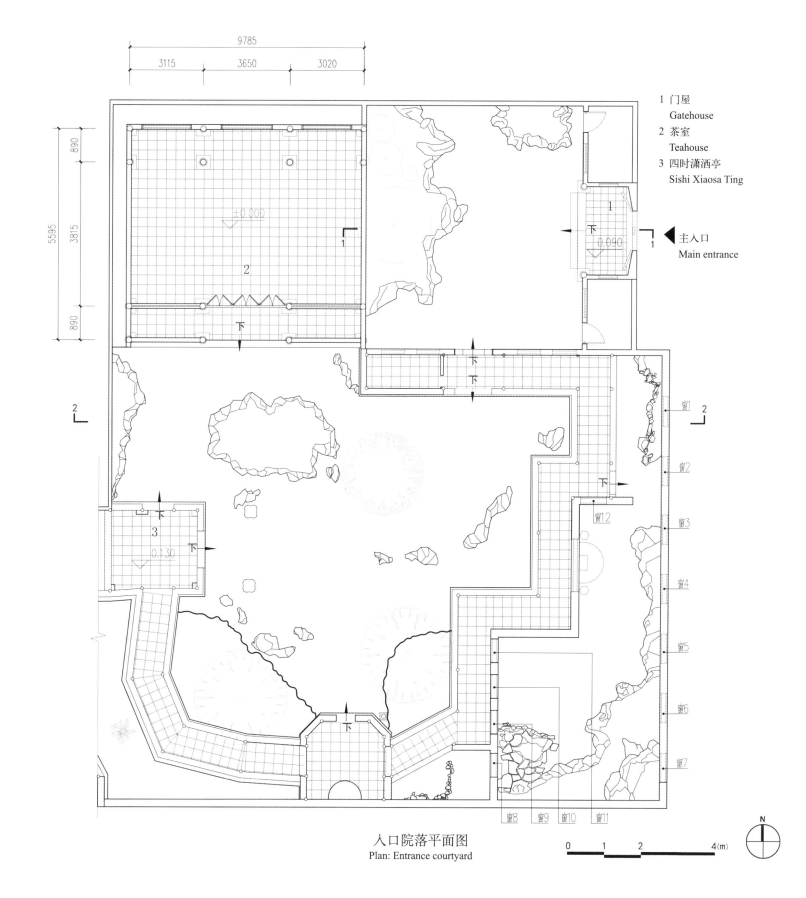

1 门屋 Gatehouse
2 茶室 Teahouse
3 四时潇洒亭 Sishi Xiaosa Ting

主入口
Main entrance

入口院落平面图
Plan: Entrance courtyard

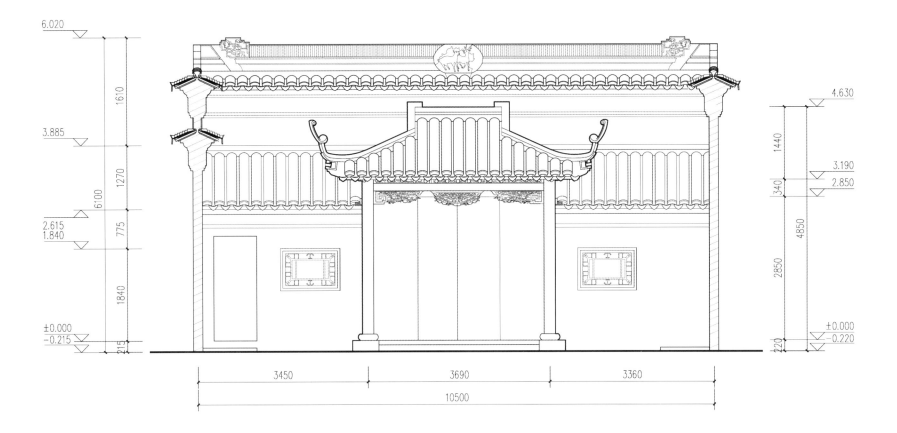

入口门屋东立面图
East elevation: Gatehouse

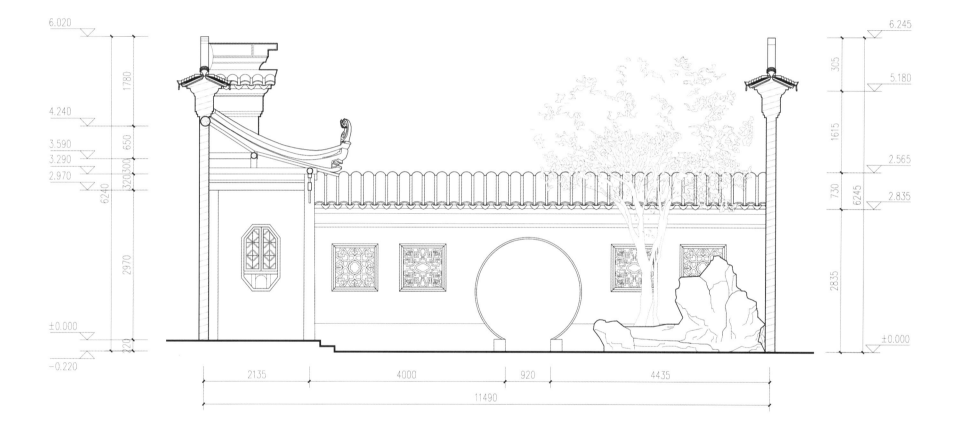

入口院落 1-1 剖面图
Section 1-1: Entrance courtyard

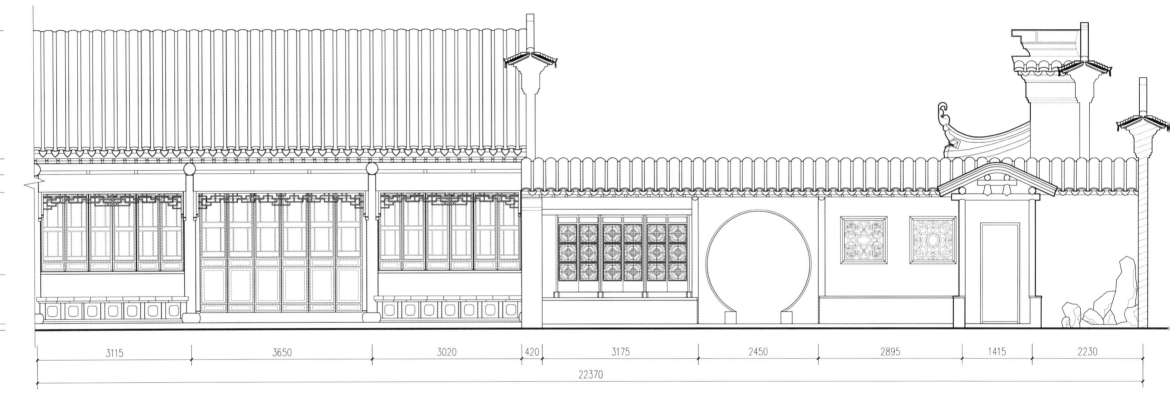

茶室
Teahouse

入口院落 2-2 剖面图
Section 2-2: Entrance courtyard

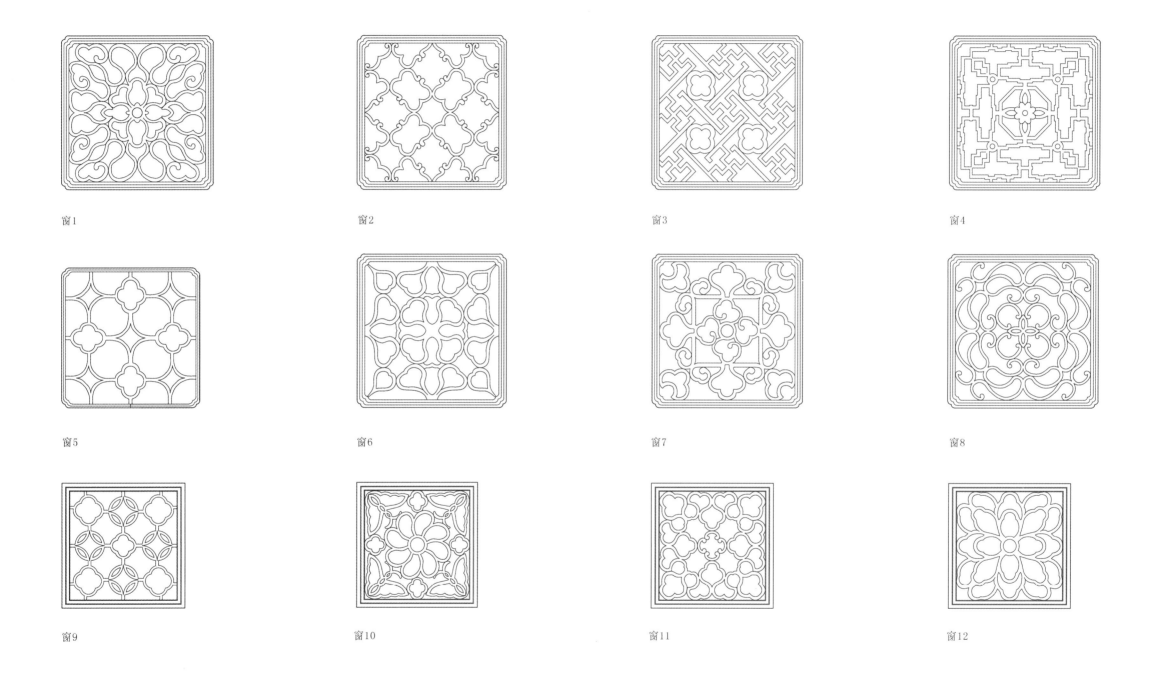

入口院落墙漏窗大样图
Details: *Louchuang* (traceried window-openings), entrance courtyard walls

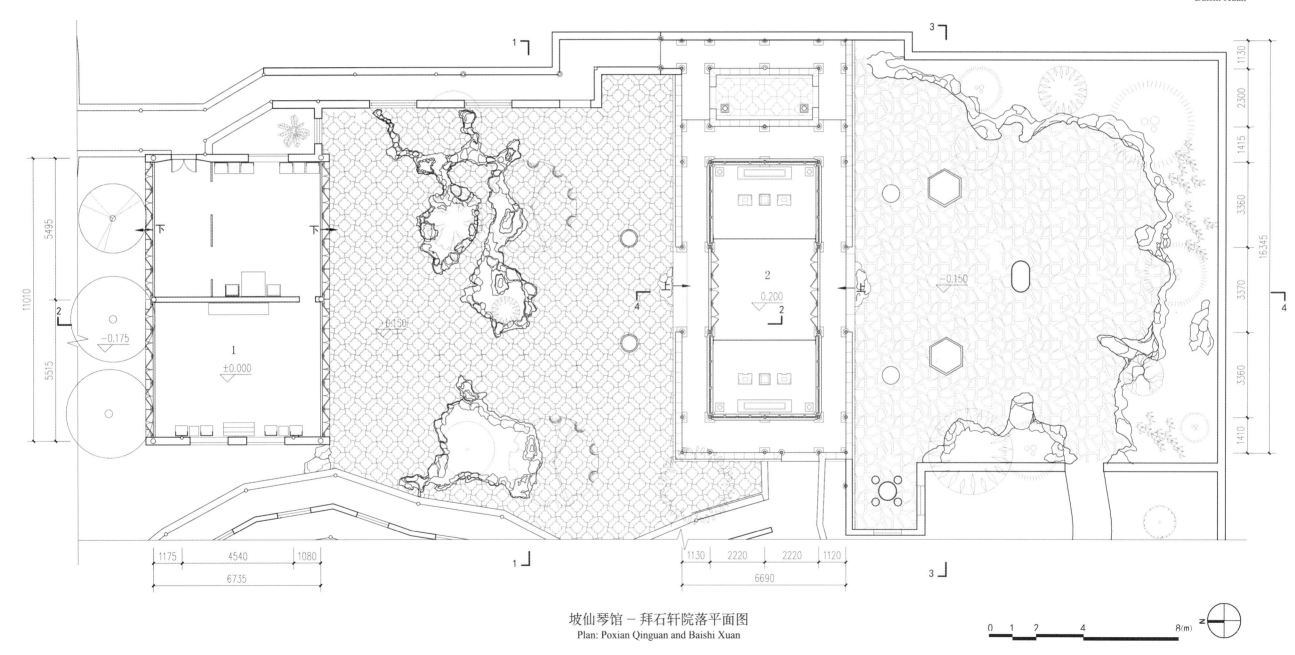

1 坡仙琴馆 Poxian Qinguan
2 拜石轩 Baishi Xuan

坡仙琴馆－拜石轩院落平面图
Plan: Poxian Qinguan and Baishi Xuan

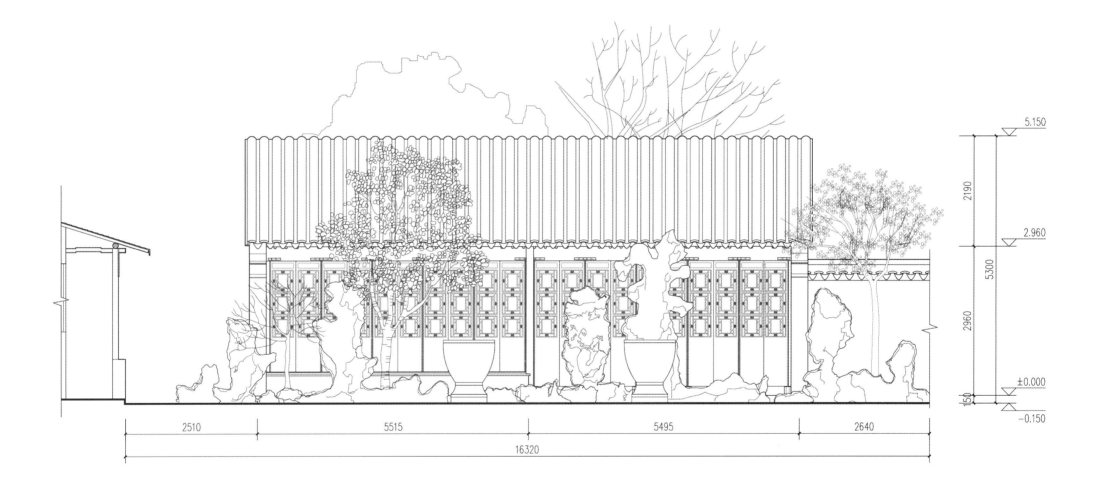

坡仙琴馆院落 1-1 剖面图
Section 1-1: Poxian Qinguan

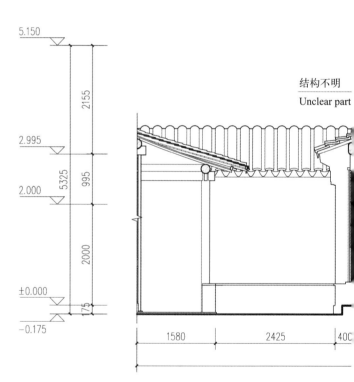

坡仙琴馆
Poxian Qinguan

坡仙琴馆院落 2-2 剖面图
Section 2-2: Poxian Qinguan

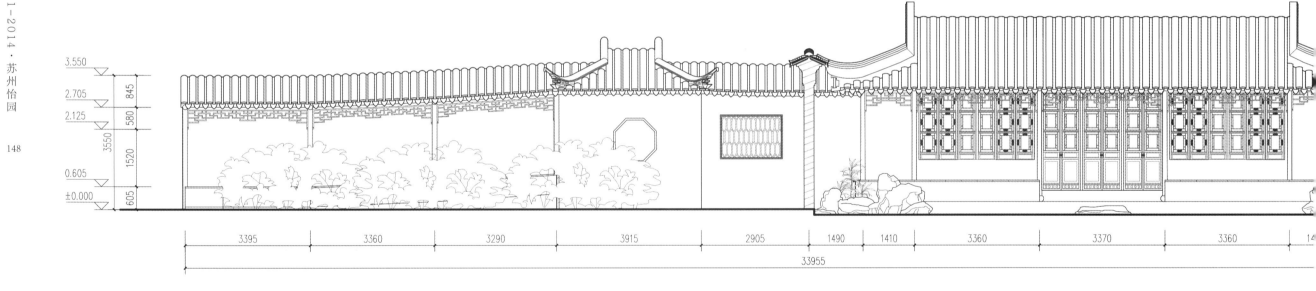

拜石轩院落 3-3 剖面图
Section 3-3: Baishi Xuan

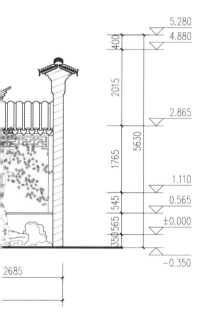

拜石轩
Baishi Xuan

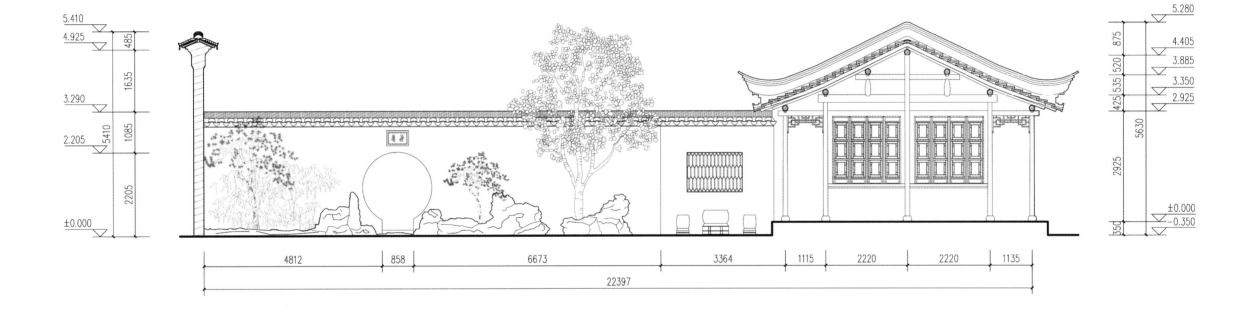

拜石轩院落 4-4 剖面图
Section 4-4: Baishi Xuan

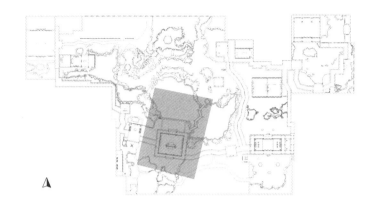

藕香榭区域
Ouxiang Xie

平板折桥及水面
The zigzag bridge and the pond

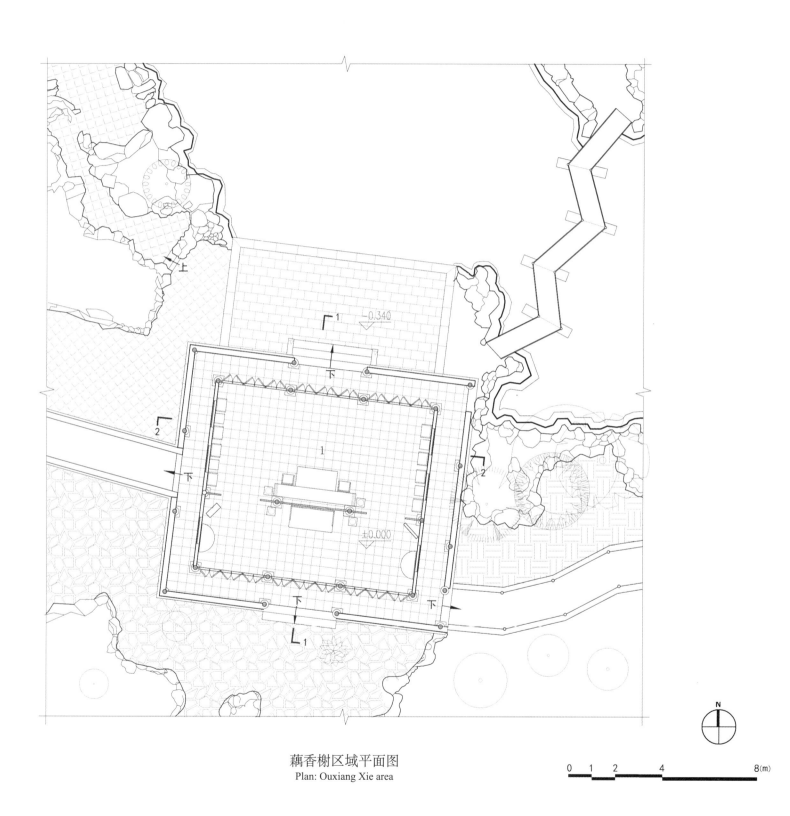

藕香榭区域平面图
Plan: Ouxiang Xie area

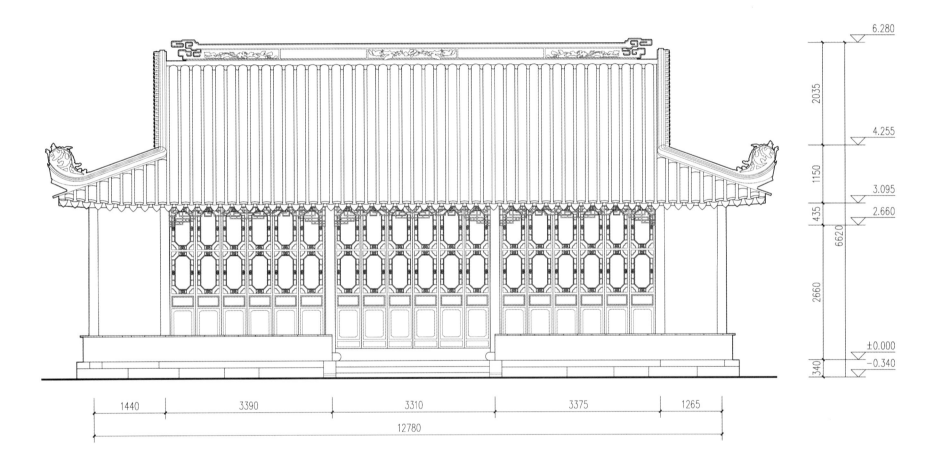

藕香榭北立面图
North elevation: Ouxiang Xie

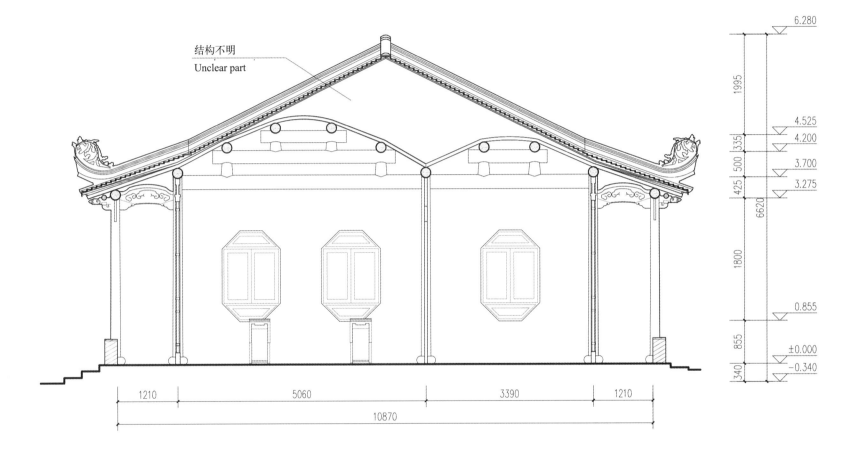

藕香榭 1-1 剖面图
Section 1-1: Ouxiang Xie

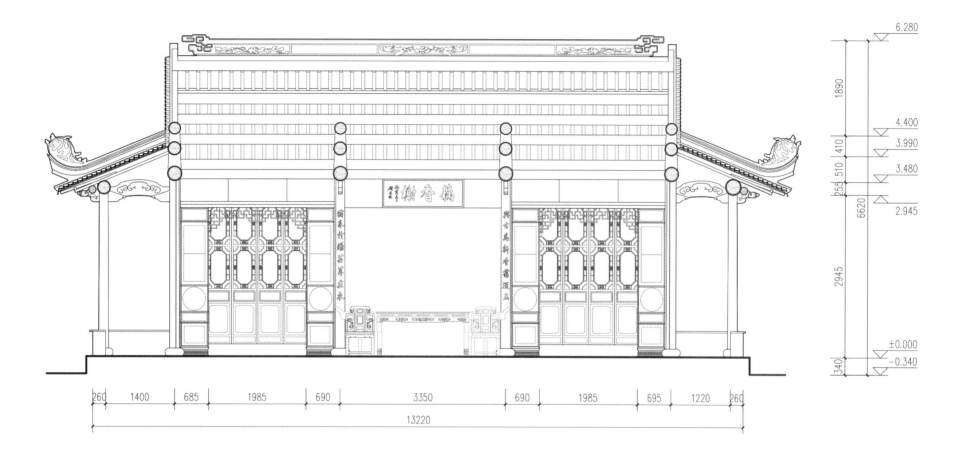

藕香榭 2-2 剖面图
Section 2-2: Ouxiang Xie

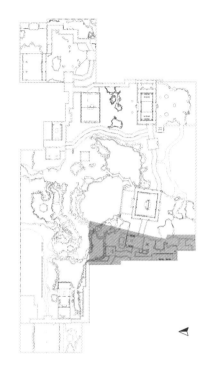

碧梧栖凤馆
Biwu Qifeng Guan

面壁亭
Mianbi Ting

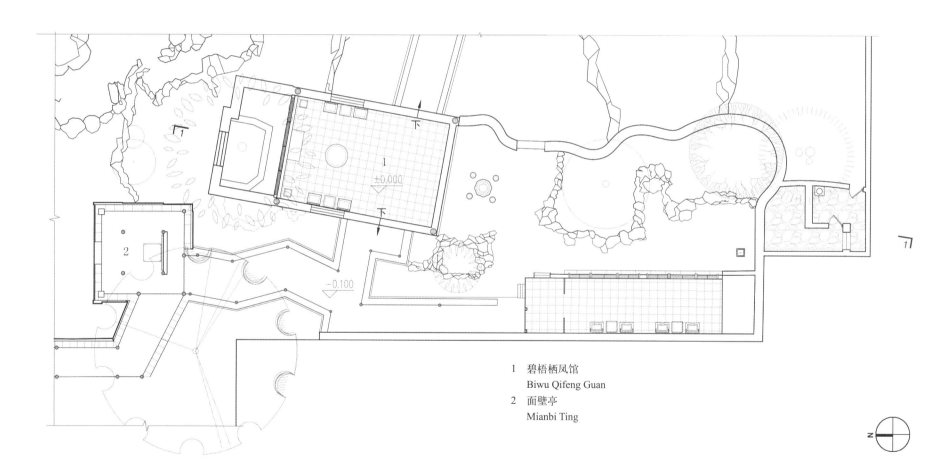

1　碧梧栖凤馆
　　Biwu Qifeng Guan
2　面壁亭
　　Mianbi Ting

碧梧栖凤馆院落平面图
Plan: Biwu Qifeng Guan

碧梧栖凤馆院落 1-1 剖面图
Section 1-1: Biwu Qifeng Guan

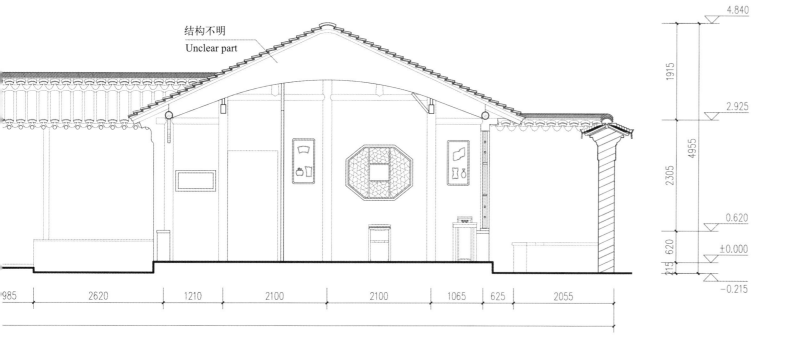

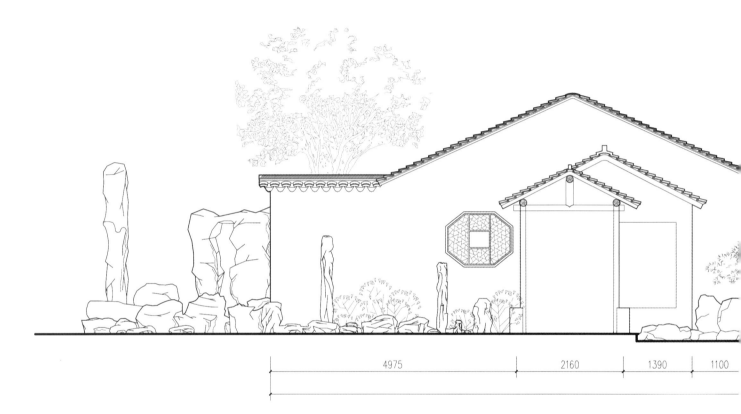

碧梧栖凤馆院落西立面图
West elevation: Biwu Qifeng Guan

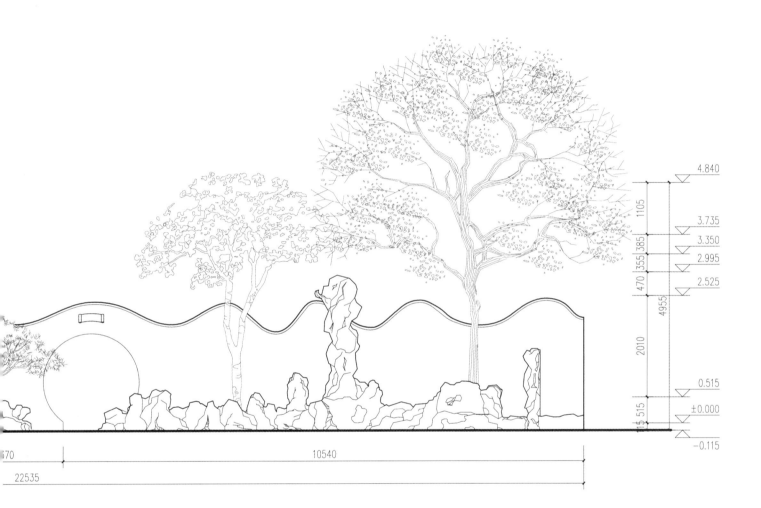

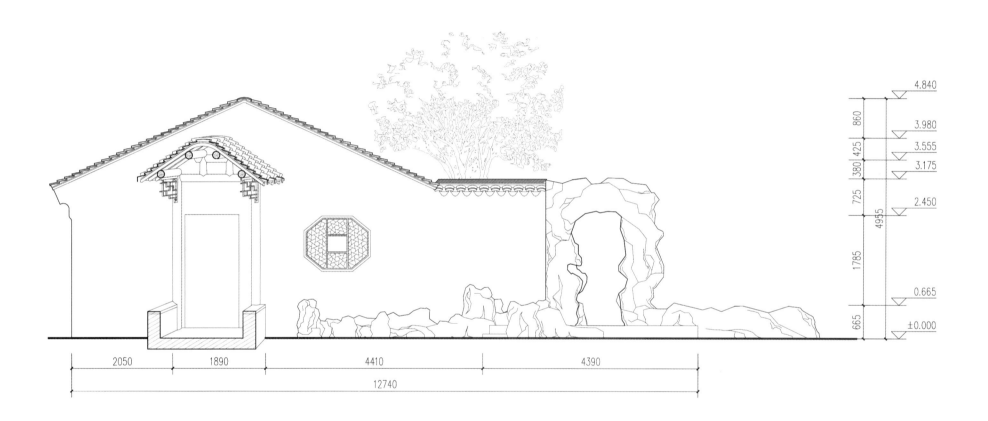

碧梧栖凤馆院落东立面图
East elevation: Biwu Qifeng Guan

面壁亭
Mianbi Ting

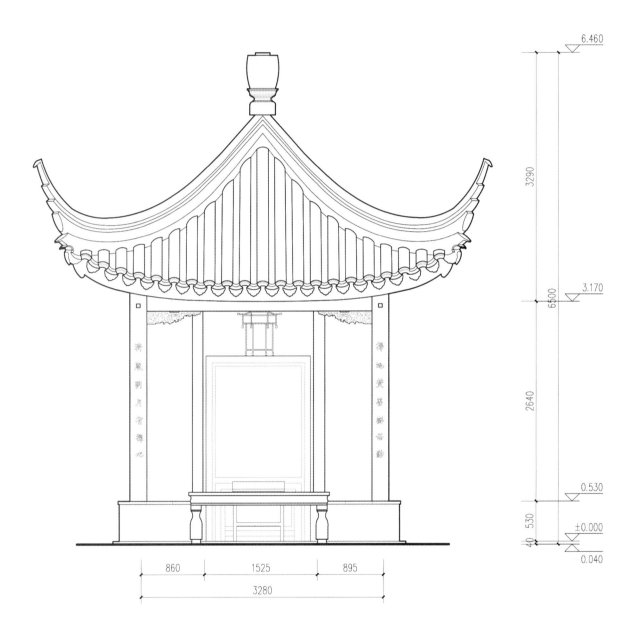

面壁亭北立面图
North elevation: Mianbi Ting

尘外画中——西安建筑科技大学古典园林测绘图辑 2011-2014·苏州怡园

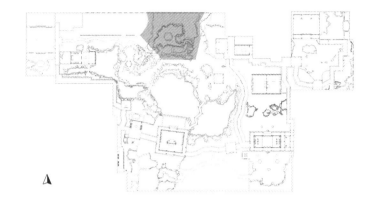

小沧浪
Xiao Canglang

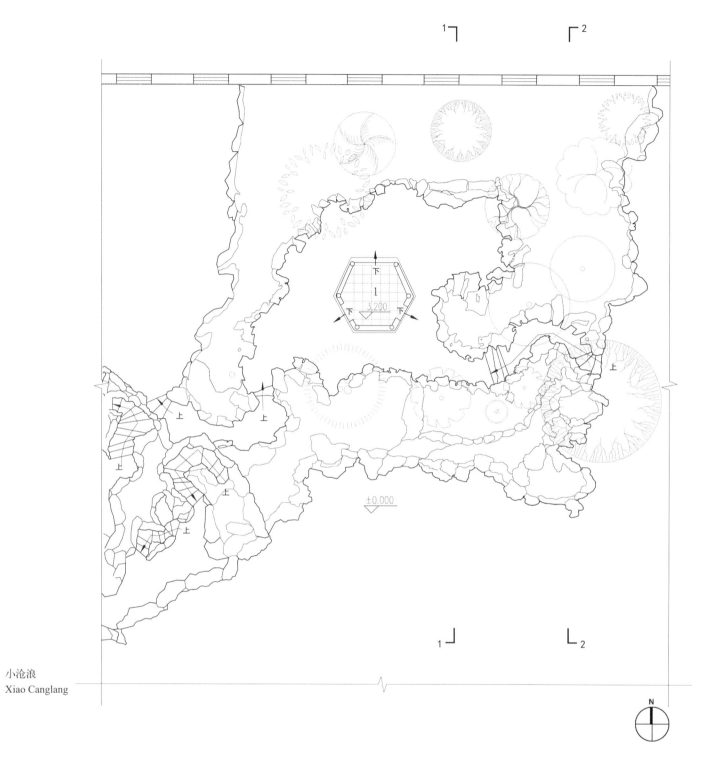

1 小沧浪
Xiao Canglang

小沧浪区域平面图
Plan: Xiao Canglang area

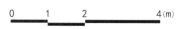

小沧浪区域 1-1 剖面图
Section 1-1: Xiao Canglang area

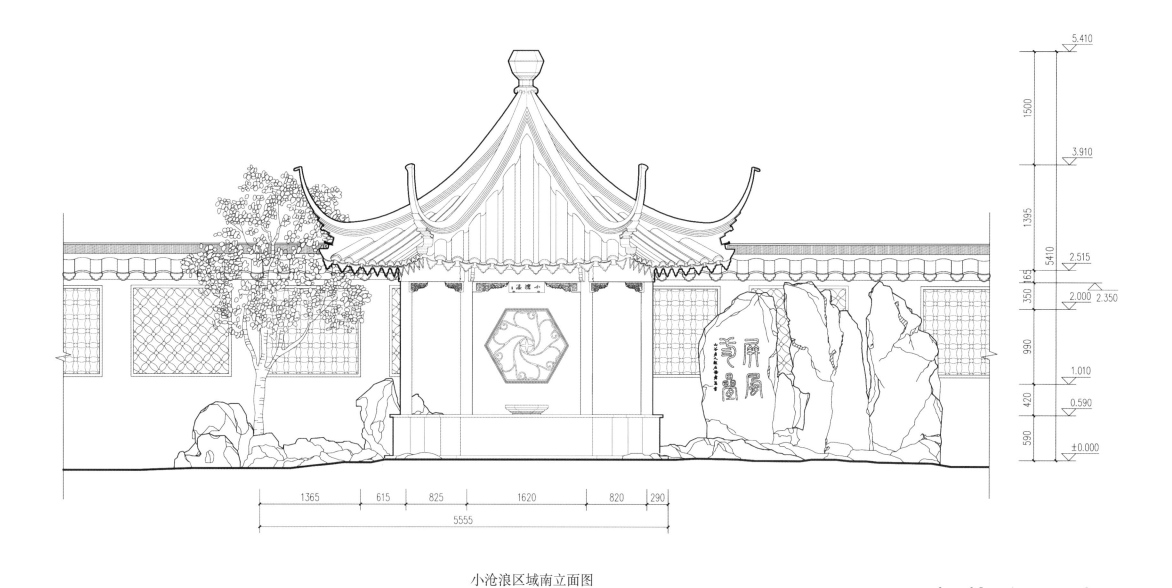

小沧浪区域南立面图
South elevation: Xiao Canglang area

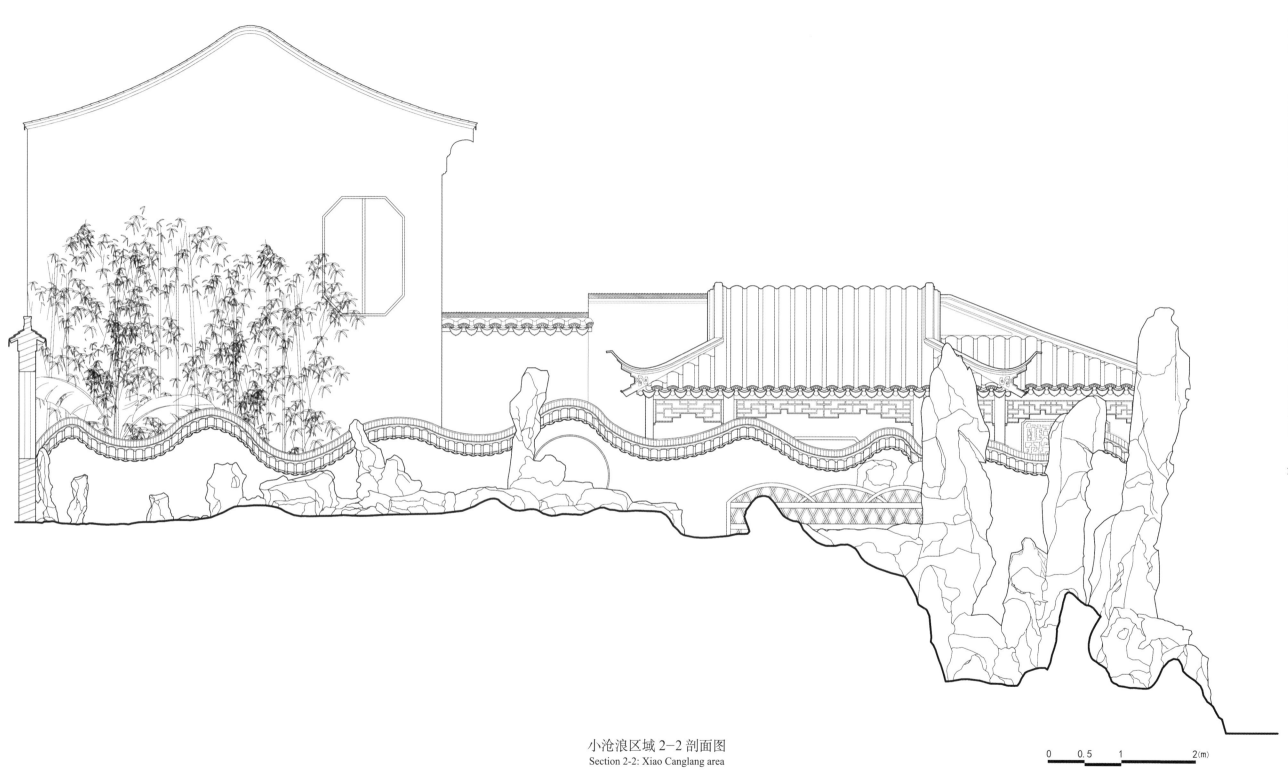

小沧浪区域 2-2 剖面图
Section 2-2: Xiao Canglang area

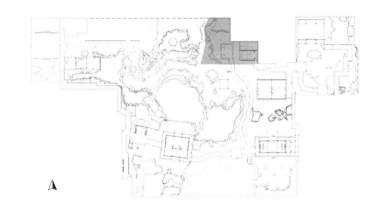

锁绿轩
Suolv Xuan

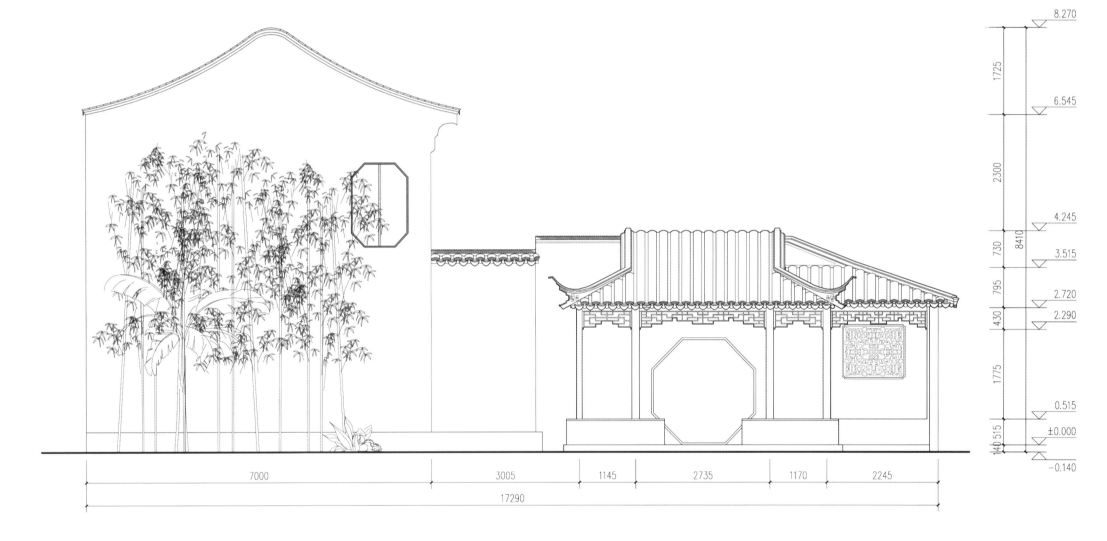

锁绿轩区域西立面图
West elevation: Suolv Xuan

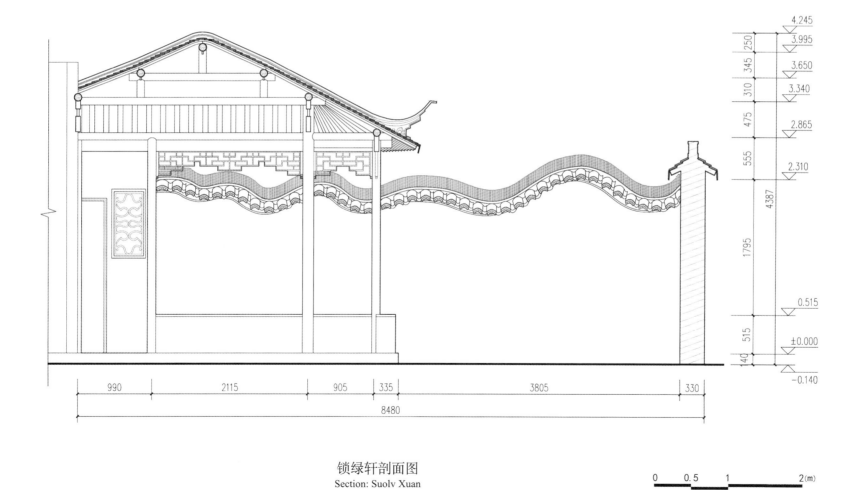

锁绿轩剖面图
Section: Suolv Xuan

螺髻亭
Luoji Ting

螺髻亭区域南立面图
South elevation: Luoji Ting

苏州沧浪亭
Canglang Ting (Canglang Pavilion), Suzhou

- 地址：苏州市老城区三元坊沧浪亭街3号
- 年代：明—清
- 保护级别：世界文化遗产，全国重点文物保护单位（第六批）

- Address: 3 Canglangting St., Sanyuan Ward, Gusu Dis., Suzhou
- Origin: 16th Century
- Status: UNESCO World Cultural Heritage Site (2000); Key Cultural Heritage Site under State Protection (2006)

05 苏州沧浪亭

苏州沧浪亭位于苏州老城城南三元坊沧浪亭街之南，西过人民路即为苏州府文庙。北宋仁宗庆历五年（1045年），退隐苏州的苏舜钦在文庙之东购得一块废地，创建了沧浪亭。苏舜钦卒后，园归章惇，章氏父子着力营建沧浪亭，使其规模和格局都发生了重大变化，时称"章园"。南宋时韩世忠（1089—1151）得到此园，将其改称"韩园"，并再次增建。韩氏之后，韩园毁于南宋末年兵乱，元时其所在地已属大云庵所有。明嘉靖间（1522—1566年）沧浪亭重建，文徵明为其题写"沧浪亭"三字楣额，归有光亦为之作《沧浪亭记》，但至明末则再度荒废。清康熙年间（1662—1722年），时任江苏巡抚的宋荦将原位于水边的亭移至水南岸土山顶上，并在园中增筑多处建筑，奠定了今日沧浪亭的基本格局。其后巡抚吴存礼、布政使梁章钜、巡抚陶澍等先后多次重修沧浪亭。咸丰年间（1851—1861年），全园再次毁于战乱，直至同治年间（1862—1875年），方由布政使应宝时与巡抚张树声主持重建，沧浪亭遂有现在所见之面貌。

沧浪亭现状总面积约16亩。园北侧临水，由平板折桥入园。园内主山居中，沧浪亭立于山顶，柱上有名联"清风明月本无价，近水远山皆有情"。入口门屋位于主山之北，其西侧一组院落为藕花水榭，东侧为面水轩与复廊；复廊蜿蜒向东直至小亭"观鱼处"；全园大部分建筑则位于主山之南，居中为明道堂和瑶华境界构成的两进院落，东为闻妙香室小院；西为清香馆、五百名贤祠及翠玲珑一组建筑。主山之西为流玉池，步碕廊和御碑亭环池之西岸。

沧浪亭与拙政园、留园、狮子林并称为"苏州四大名园"。其于1982年列为江苏省文物保护单位，2000年作为"苏州古典园林"的增补项目列入《世界遗产名录》，2006年提升为第六批全国重点文物保护单位。

图1：临水复廊
Fig.1: Waterside *fulang*

主要参考文献：

[1] [宋] 苏舜钦. 苏舜钦集编年校注 [M]. 傅平骧，胡问涛. 校注. 北京：中华书局，1962.

[2] [宋] 叶梦得. 石林诗话校注 [M]. 逯铭昕. 校注. 北京：人民文学出版社，2012.

[3] [明] 林世远，王鏊. 正德姑苏志 [M]. (北京图书馆古籍珍本丛刊·第二十六至二十七). 明正德刻嘉靖续修本影印. 六十卷. 北京：书目文献出版社，2000.

[4] [明] 卢熊. 苏州府志 [M]. (中国方志丛书·华中地方·第432号). 据洪武十二年钞本影印. 五十卷图一卷. 台北：成文出版社，1983.

[5] [清] 李铭皖，冯桂芬. 同治苏州府志 [M]. (中国地方志集成·江苏府县志辑·第七至十册). 据光绪九年刻本影印. 一百五十卷首三卷. 南京：江苏古籍出版社，1991.

[6] [清] 宋荦. 沧浪小志 [M]. (故宫珍本丛刊·239册). 据光绪十年刻本影印. 二卷. 海口：海南出版社，2001：489-518 页.

[7] 曹允源、李根源. 民国吴县志 [M]. (中国地方志集成·江苏府县志辑·第十一至十二册). 据民国二十二年苏州文新公司铅印本影印. 八十卷. 江苏古籍出版社，1991.

[8] [清] 徐崧、张大纯. 百城烟水 [M]. 薛正兴. 校点. 南京：江苏古籍出版社，1999.

测 绘：

西安建筑科技大学建筑学院风景园林学 2011 级 01，02 班——

黄莹、李伊婷、兰帆、郝晟、刘明；张月、云鹤、冉艺辉、李萌、李鑫、蔡雨彤；白皓月、刘雯西、李蔓、陈岩、畅茹茜、段优；常昊翀、都凯、李琼、连萌、杨洁琼、庄晓眉；王国今、王霄、刘婉莹、贺小峰、常禾、常青；邵佳慧、吴昕泽、刘昱、周琦、张欢；杨烜子、宋茜伦、张国帅、陈璐、秦荣利；孙易翀、王浩、董文煊、刘媛、杨瑾、潘晓佳；张琳琳、张超、杨培培、武儒、赵安妮； 綦琪、仇静、马晶楠、王珂、张泽豪；

建筑历史与理论、建筑设计研究生 2014 级——

李双双、李宛儒、李祯、王琳、蔡楠、王端正、李路斌；王泽鑫、王忱、王彧萱、孙国才、严骏、陈冠宇、王桦；

指导教师：林源、岳岩敏、喻梦哲；
图纸绘制、整理：汶武娟、马英晨；
测绘时间：2014 年 10 月

图 2：沧浪亭入口
Fig.2: Entrance

图 4：面水轩
Fig.4: Mianshui Xuan

图 3：沧浪亭
Fig.3: Canglang Ting

图 5：流玉潭
Fig.5: Liuyu Tan

图 6：明道堂
Fig.6: Mingdao Tang

图 7：翠玲珑
Fig.7: Cui Linglong

05. Canglang Ting (Canglang Pavilion)

Named after the most prominent garden structure within its perimeters, the classical garden of the Canglang Pavilion is located on Canglangting St., Sanyuan Ward, in southern historic urban area of Suzhou and near the city's Confucian Temple to its west across Renmin Rd. In 1045, the renowned Northern Song (960–1127) literati SU Shunqing (1008–1048) purchased, upon retirement, an anteriorly unoccupied stretch of land east to the Confucian Temple to originally build the garden. After his death, the site came to be owned by the dignified politician ZHANG Dun (1035–1105). ZHANG and his son put effort into the garden, which significantly transfigured its layout and size. A new name, ZHANG Garden, was also assigned, until it found its way in Southern Song period (1127–1279) into the possession of HAN Shizhong (1089–1151), an eminent general. Like his predecessors HAN also expanded the place and renamed it to HAN Garden. However, soon after HANs' days chaotic warfare brought down the garden along with the Song Empire. During Yuan (1271–1368) and early Ming Dynasties (1368–1644), the site was among the properties of Dayun An (Grand-Cloud Buddhist Nunnery), until reconstruction of the garden during the reign of Emperor Shizong of Ming (r. 1522–1566). WEN Zhengming (1470–1559), a prominent calligrapher and painter, offered his handwriting for the name plaque and GUI Youguang (1507–1571), a renowned literati indited the essay *Canglang Ting Ji* (On Canglang Pavilion) to commemorate the restored grandeur. However, having failed again to avoid the fate of decline in the last days of Ming Empire, the garden had to become subject to another round of restoration in early Qing period (1644–1912). SONG Luo (1634–1714), then the prefect of Jiangsu Province and the commissioner of the project, relocated the naming pavilion from its former waterside location to the top of the rockery hill on south pond bank and added various new structures within the garden's perimeters. The very layout of the site we see today is largely Song's legacy. Following Song, high local officials such as WU Cunli, LIANG Zhangju (1775–1849) and TAO Shu (1779–1839) successively ordered projects of repair for the Garden, only to lose it once again to the chaos of armed conflicts in 1850s. In 1872, Governor YING Baoshi (1821–1890) and Prefect ZHANG Shusheng (1824–1884) launched works of reconstruction once again, which finished the final touch on the garden's present-day appearance.

The garden now covers approximately 1.07 ha. Its north border abuts on water, over which a zigzag stone-slab bridge leads to the main entrance. The site is dominated by a rockery hill, named Zhen Shanlin (Authentic Landscape), at the center. On the hill stands the eponymous pavilion bearing the antithetical couplet inscribed on tablets that reads "Qing-Feng Ming-Yue Ben Wu-Jia, Jin-Shan Yuan-Shui Jie You-Qing" (roughly meaning that natural beauties are not monetarily but sentimentally valuable to humans). North to the hill stands the entrance gatehouse. From the entrance one could walk westward to the courtyards featuring Ouhua Shuixie (Lotus-Root Waterside Pavilion) or eastward to Mianshui Xuan (Water-Facing Pavilion) and a set of *fulang* (both-side-opening galleries), which lead sinuously further eastward to Guanyu Chu (Fish-Viewing Pavilion). However, most garden structures are located south of the hill, which could be divided to three groups. The east group comprises of Wen Miaoxiang Shi (Mysterious-Incense-Smelling Chamber), the middle the courtyards of Ming-dao Tang (Bright-Path Hall) and Yaohua Jingjie (Heavenly Realm), and the west Qingxiang Guan (Subtle-Fragrance House), Wubai Mingxian Ci (500 Sages' Shrine) and Cui Linglong (Turquoise-Exquisiteness Pavilion). Additionally, to the west of the hill stands Yu Beiting (Imperial Stele-Pavilion) and Buqi Lang (Arduous-Journey Galleries). The pond, Liuyu Tan (Flowing-Jade Pond), lies between the hill and Buqi Lang.

The Canglang Pavilion enjoys a dignified reputation, as known as one of 'The Four Renowned Classical Gardens of Suzhou', of which the other three being Zhuozheng Yuan (Humble Administrator's Garden), Liu Yuan (Lingering Garden) and Shizi Lin (Lion Grove Garden). It was listed as a Cultural Heritage Site under Provincial Protection in 1982 and a Key Cultural Heritage Site under State Protection in 2006 (6th Round of Nomination). Also in 2000 it joined, as a part of the extension, the inscribed UNESCO World Cultural Heritage Site of *the Classical Gardens of Suzhou*.

Bibliography

[1] FU P. and HU, W. (1962) *Su Shunqing Ji Biannian Jiaozhu* (A Collation and Annotations for the Chronologically Arranged Works of Su Shunqing). Beijing: Zhonghua Book Company (in traditional Chinese).

[2] YE, M. (2012) *Shilin Shihua Jiaozhu* (A Collation and Annotations for *Shi-Lin's Notes on Poetry*). Beijing: People's Literature Press (in traditional Chinese).

[3] LIN,S. and WANG, A. (2000) *Gusu Zhi* (Records of Gusu). Beijing: Zhonghua Book Company (in traditional Chinese).

[4] LU, X. (1983) *Suzhou Fu Zhi* (1368-1398 Complied Records of Suzhou). Yangzhou: Guangling Press (in traditional Chinese).

[5] LI, M. and FENG, G. (1991) *Tongzhi Suzhou Fu Zhi* (1862–1874 Compiled Records of Suzhou). Nanjing: Jiangsu Guji Press (in traditional Chinese).

[6] SONG, L. (2001) *Canglang Xiao Zhi* (Trivial Records of Canglang Pavilion). Haikou: Hainan Press (in traditional Chinese).

[7] CAO, Y. and LI, G. (1991) *Minguo Wuxian Zhi* (ROC Compiled Records for the County of Wu). Nanjing: Jiangsu Guji Press (in traditional Chinese).

图 8：仰止亭
Fig.8: Yangzhi Ting

[8] XU, S. and ZHANG, D. (1999) *Baicheng Yanshui* (Attractions of 100 Towns). Nanjing: Jiangsu Guji Press (in traditional Chinese).

Credits

Surveyors:

2016 Undergraduate Class A & B of Landscape Architecture (in groups):

HUANG Ying, LI Yiting, LAN Fan, HAO Sheng, LIU Ming; ZHANG Yue, YUN He, RAN Yihui, LI Meng, LI Xin, CAI Yutong; BAI Haoyue, LIU Wenxi, LI Man, CHEN Yan, CHANG Ruqian, DUAN You; CHANG Haochong, DU Kai, LI Qiong, LIAN Meng, YANG Jieqiong, ZHUANG Xiaomei; WANG Guojin, WANG Xiao, LIU Wanying, HE Xiaofeng, CHANG He, CHANG Qing; SHAO Jiahui, WU Xinze, LIU Yu, ZHOU Qi, ZHANG Huan; YANG Xuanzi, SONG Qianlun, ZHANG Guoshuai, CHEN Lu, QIN Rongli; SUN Yichong, WANG Hao, DONG Wenxuan, LIU Yuan, YANG Jin, PAN Xiaojia; ZHANG Linlin, ZHANG Chao, YANG Peipei, WU Ru, ZHAO Anni; QI Qi, QIU Jing, MA Jingnan, WANG Ke, ZHANG Zehao

2017 Postgraduate Classes of Architectural Design and Architectural History:

LI Shuangshuang, LI Wanru, LI Zhen, WANG Lin, CAI Nan, WANG Duanzheng, Li Lubin, WANG Zexin, WANG Chen, WANG Yuxuan, SUN Guocai, YAN Jun, CHEN Guanyu, WANG Hua

Tutors: LIN Yuan, YUE Yanmin, YU Mengzhe

Drawing Preparation: WEN Wujuan, MA Yingchen

Date: October, 2014

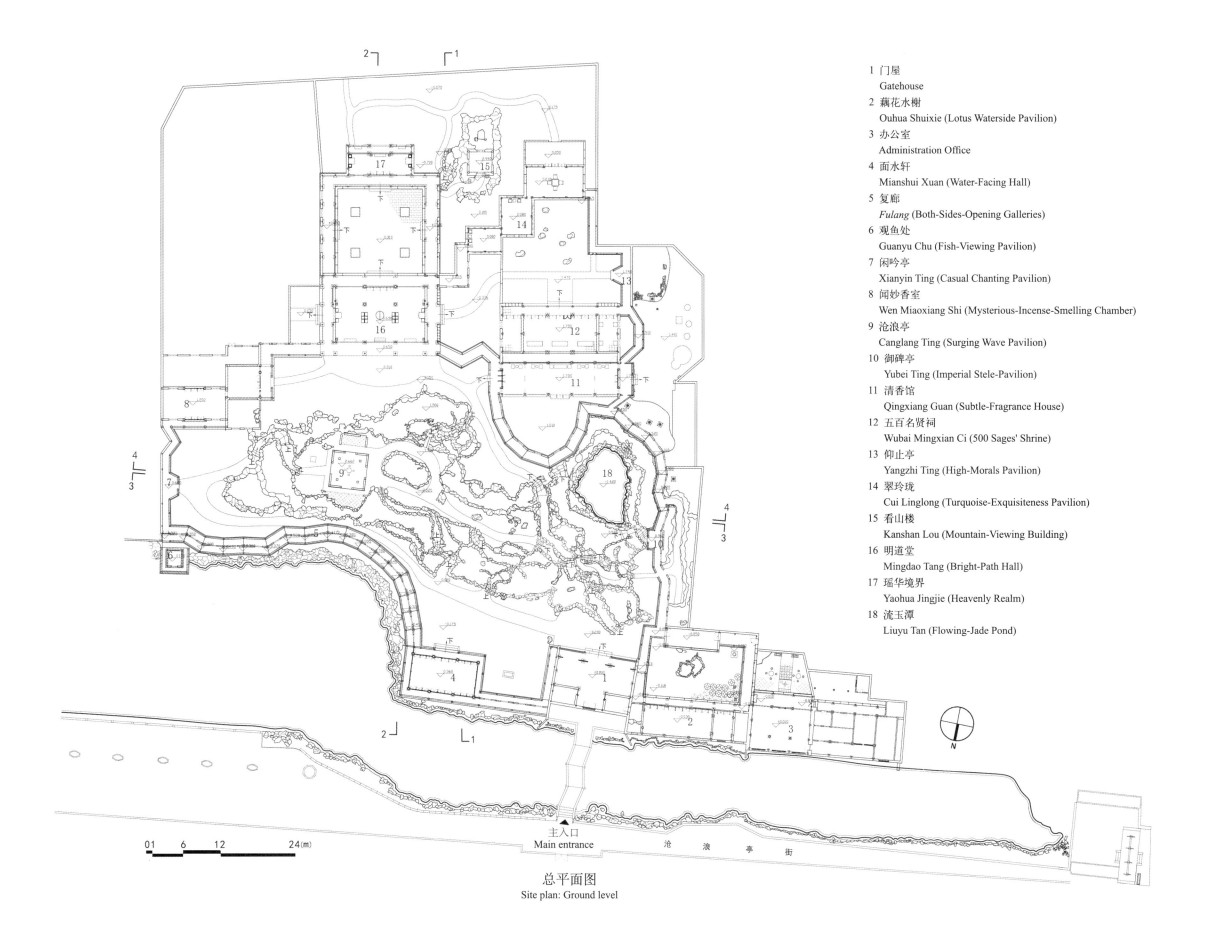

总平面图
Site plan: Ground level

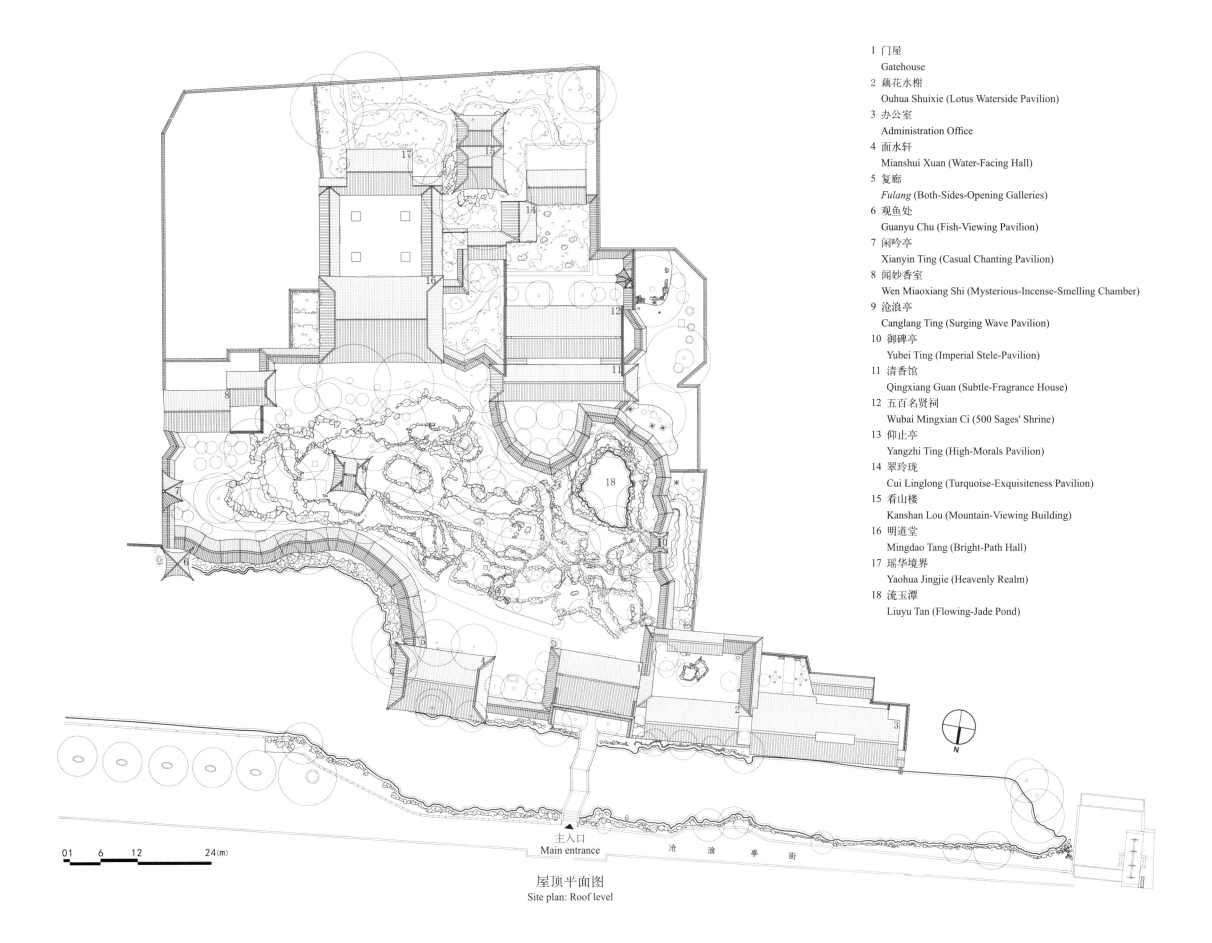

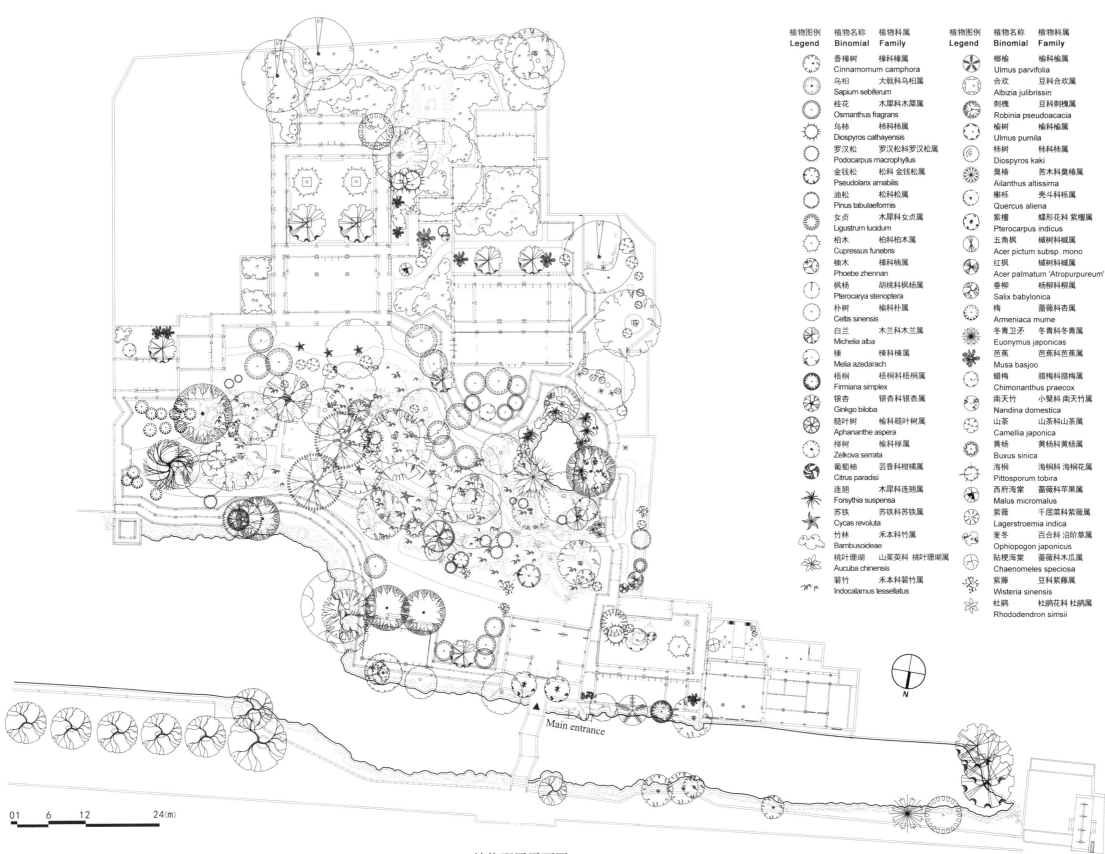

植物配置平面图
Vegetation

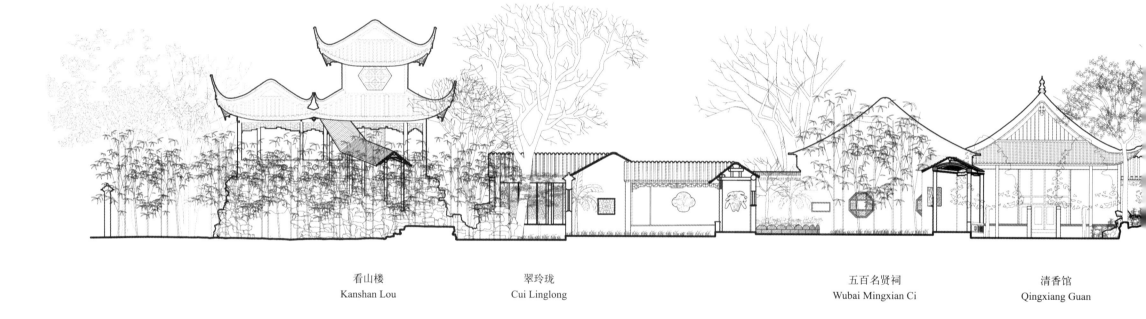

| 看山楼 | 翠玲珑 | 五百名贤祠 | 清香馆 |
| Kanshan Lou | Cui Linglong | Wubai Mingxian Ci | Qingxiang Guan |

全园 1-1 剖面图
Section 1-1: The whole garden

御碑亭
Yubei Ting

面水轩
Mianshui Xuan

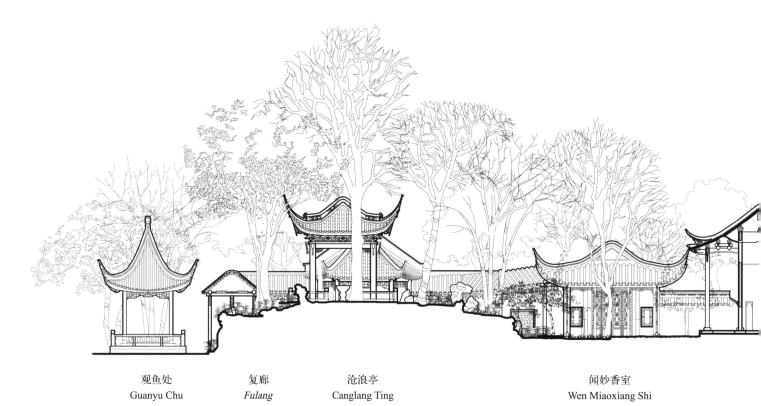

| 观鱼处 | 复廊 | 沧浪亭 | 闻妙香室 |
| Guanyu Chu | Fulang | Canglang Ting | Wen Miaoxiang Shi |

全园 2-2 剖面图
Section 2-2: The whole garden

明道堂　　　　　　　　　　　瑶华境界
Mingdao Tang　　　　　　　Yaohua Jingjie

| 御碑亭 | 流玉潭 | 面水轩 |
| Yubei Ting | Liuyu Tan | Mianshui Xuan |

全园 3-3 剖面图（北视）
Section 3-3 (West-East): The whole garden

闲吟亭
Xianyin Ting

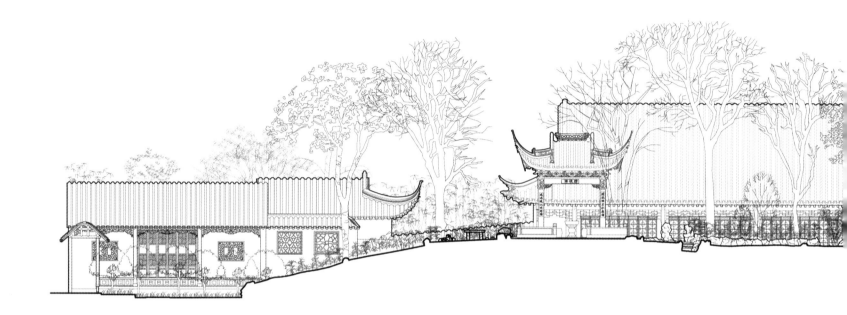

| 闻妙香室 | 沧浪亭 | 明道堂 |
| Wen Miaoxiang Shi | Canglang Ting | Mingdao Tang |

全园 3-3 剖面图（南视）
Section 3-3 (East-West): The whole garden

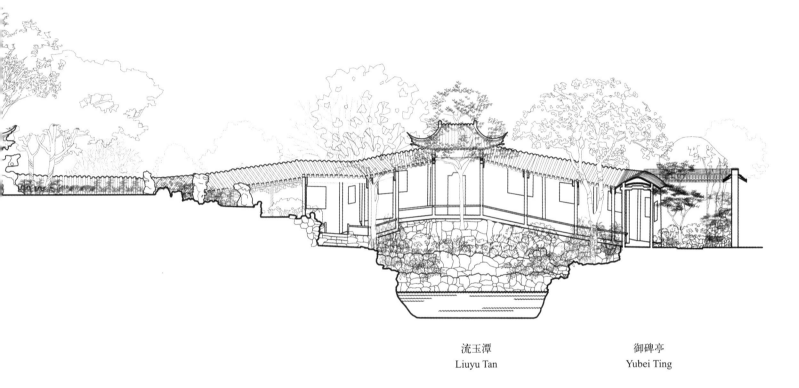

流玉潭　　　　　御碑亭
Liuyu Tan　　　Yubei Ting

| 观鱼处 | 复廊 | 面水轩 |
| Guanyu Chu | Fulang | Mianshui Xuan |

全园沿河北立面图

North elevation: The whole garden (riverside)

藕花水榭
Ouhua Shuixie

尘外画中——西安建筑科技大学古典园林测绘图辑 2011—2014·苏州沧浪亭

沧浪亭
Canglang Ting

全园真山林区域北部展开立面图
Flattened elevation: Northern part, Zhen Shanlin area

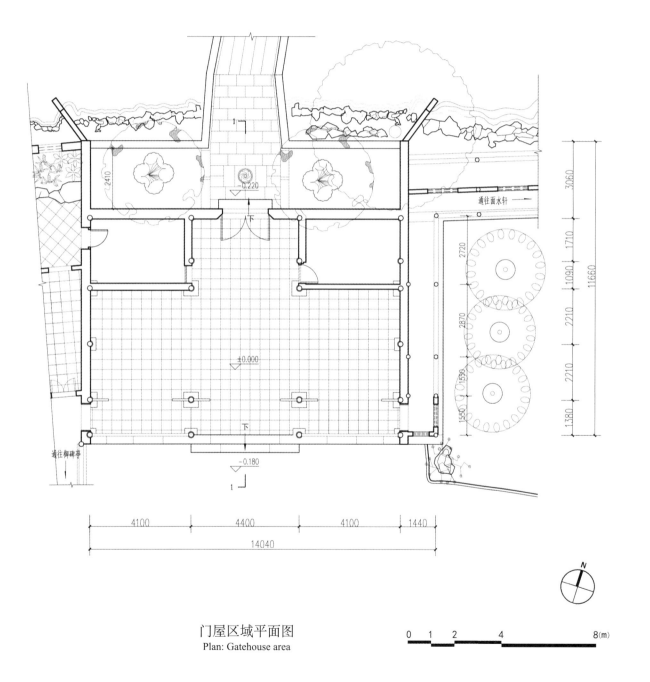
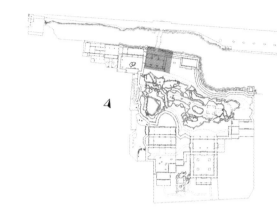
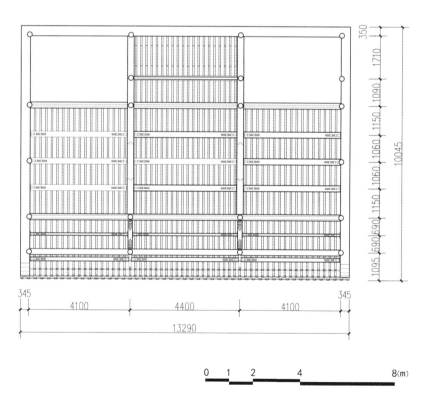

门屋区域平面图
Plan: Gatehouse area

门屋梁架仰视图
Reflected truss plan: Gatehouse

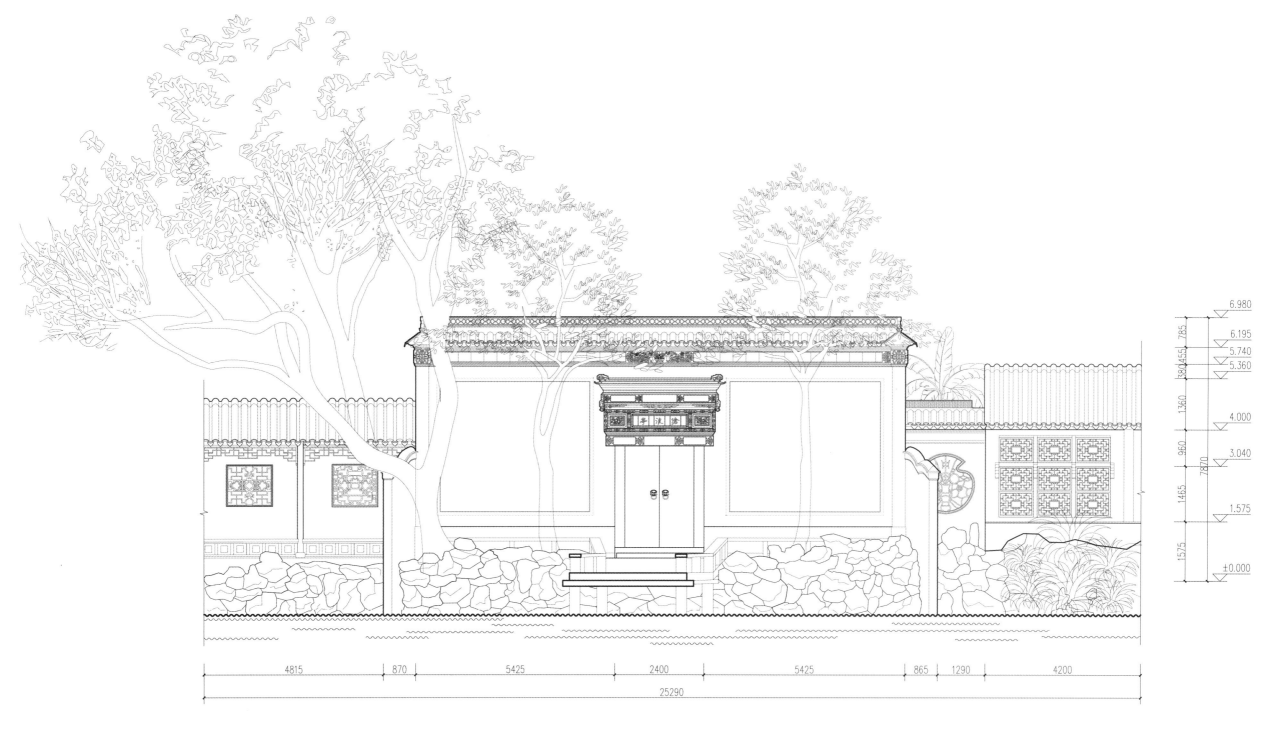

入口区域北立面图
North elevation: Entrance area

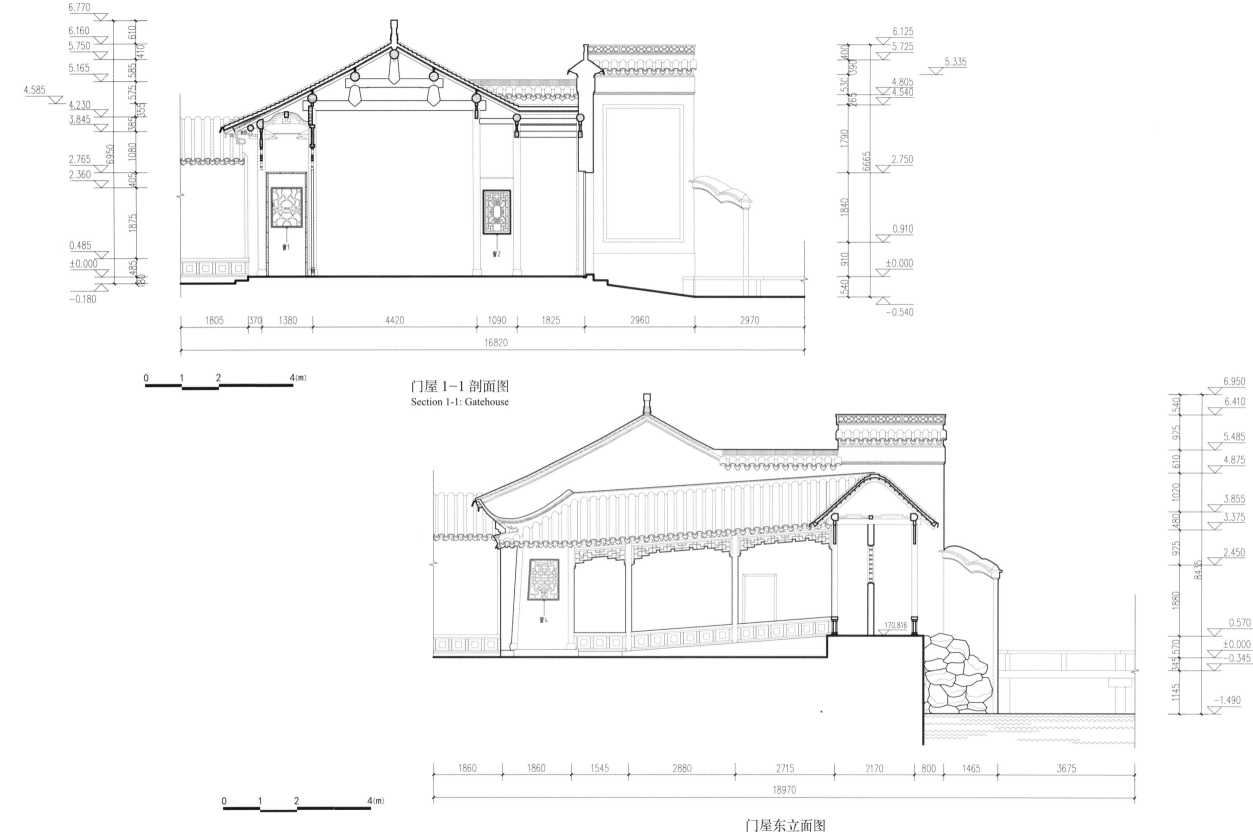

门屋 1-1 剖面图
Section 1-1: Gatehouse

门屋东立面图
East elevation: Gatehouse

窗1

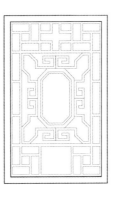

窗2

窗3

窗4

门屋区域漏窗大样图

Details: *Louchuang* (traceried window-openings), Gatehouse area

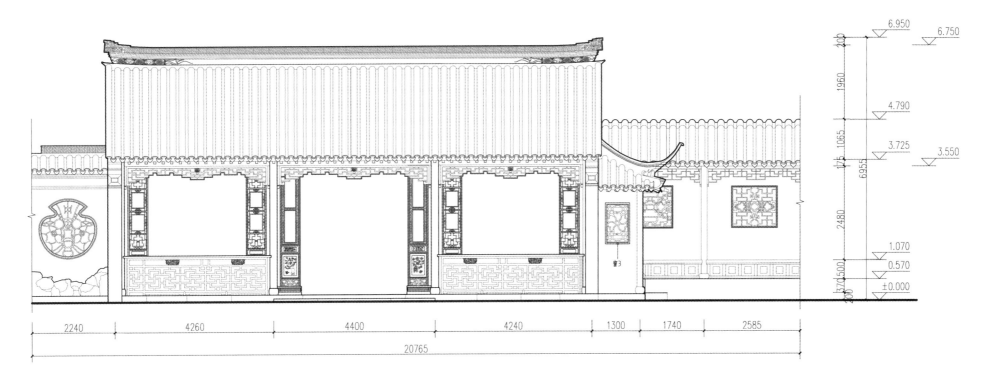

门屋南立面图

South elevation: Gatehouse

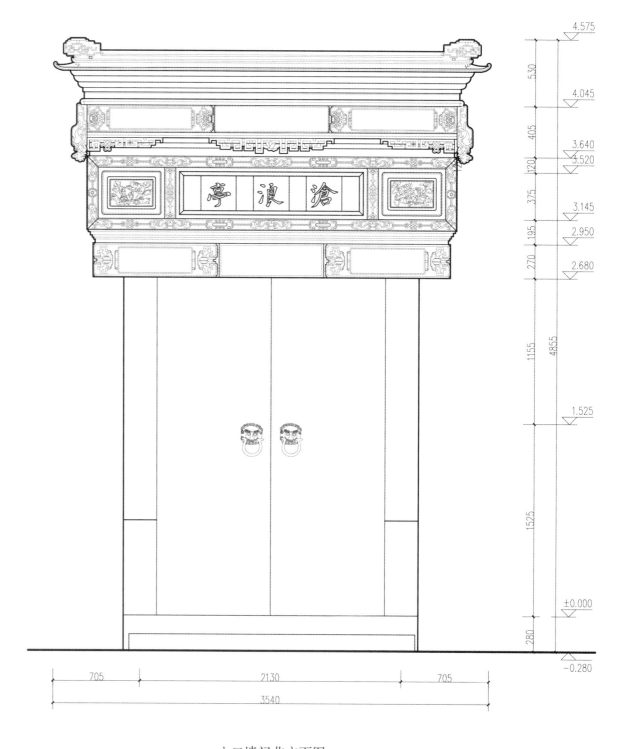

入口墙门北立面图
North elevation: Gatehouse

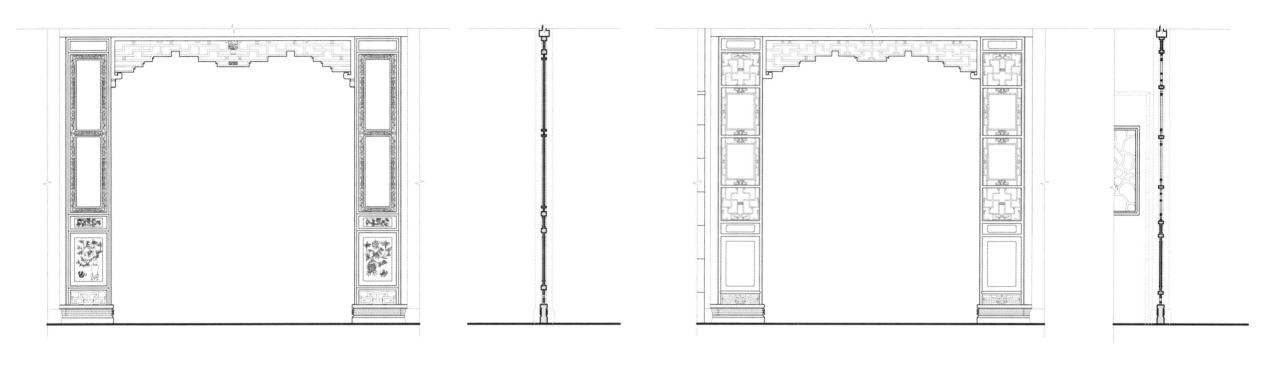

门屋当心间落地罩大样图
Details: *Luodizhao* (traceried ceiling-to-floor screen), central bay, Gatehouse

门屋次间落地罩大样图
Details: *Luodizhao*, secondary bay, Gatehouse

尘外画中——西安建筑科技大学古典园林测绘图辑 2011-2014·苏州沧浪亭

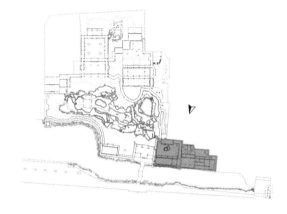

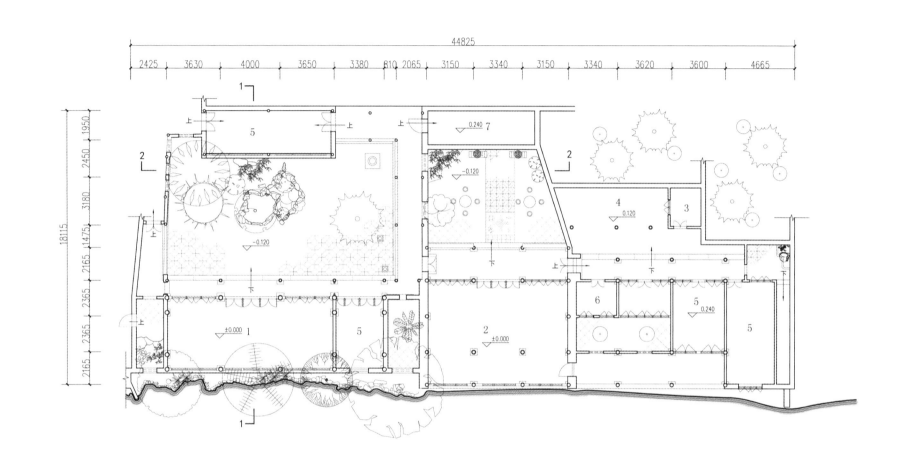

1 藕花水榭
  Ouhua Shuixie
2 锄月轩
  Chuyue Xuan
3 杂物间
  Utility room
4 车棚
  Bicycle shed
5 办公室
  Office
6 卫生间
  Toilet
7 开水房
  Boiler room

藕花水榭及办公区平面图
Plan: Ouhua Shuixie and office area

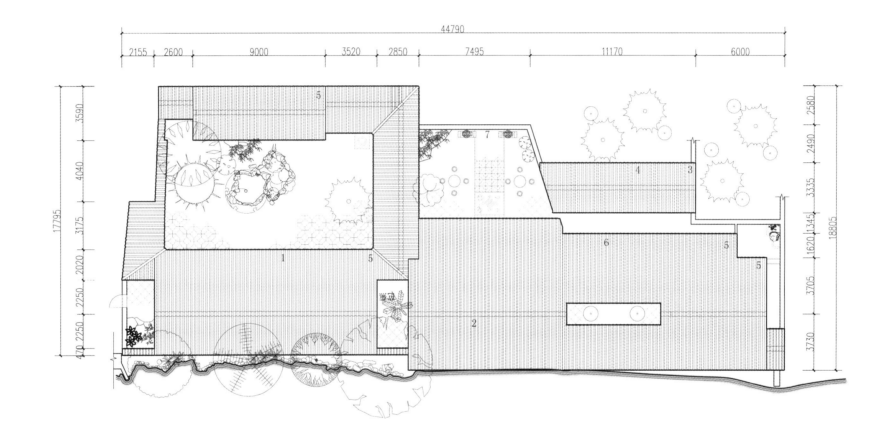

1 藕花水榭
  Ouhua Shuixie
2 锄月轩
  Chuyue Xuan
3 杂物间
  Utility room
4 车棚
  Bicycle shed
5 办公室
  Office
6 卫生间
  Toilet
7 开水房
  Boiler room

藕花水榭及办公区屋顶平面图
Roof plan: Ouhua Shuixie and office area

窗 1—窗 2 大样图
Details: Windows nos.1&2

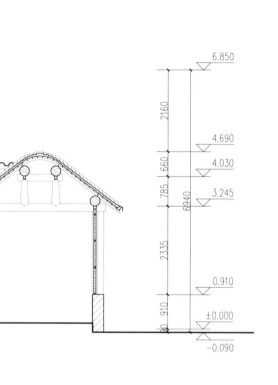

藕花水榭院落 1-1 剖面图
Section 1-1: Ouhua Shuixie

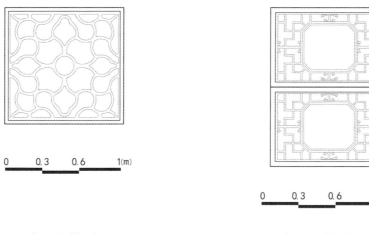

窗 1 大样图
Details: Window no.1

窗 2 大样图
Details: Window no.2

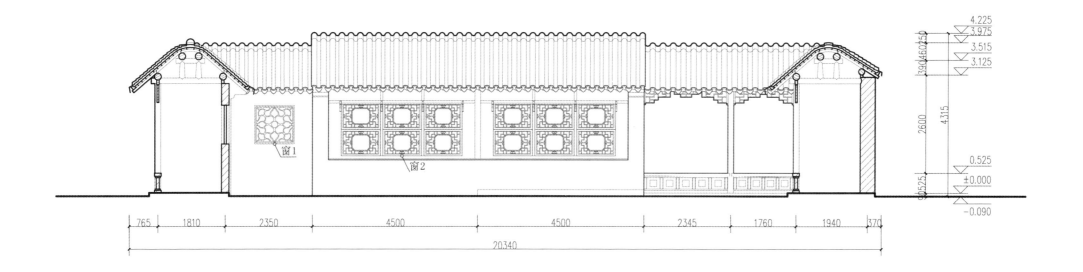

藕花水榭院落 2-2 剖面图
Section 2-2: Ouhua Shuixie

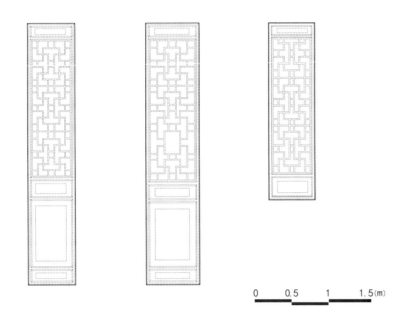

格扇1、格扇2、窗3大样图
Details: *Geshan* (partition doors) nos.1-2 and window no. 3

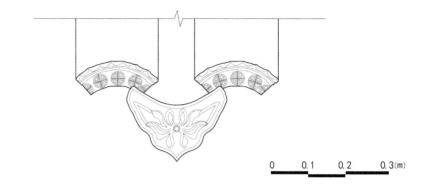

花边瓦－滴水大样图
Details: *Huabinwa* (petal-shaped-edge tiles) and a *Dishui* (flashing tile)

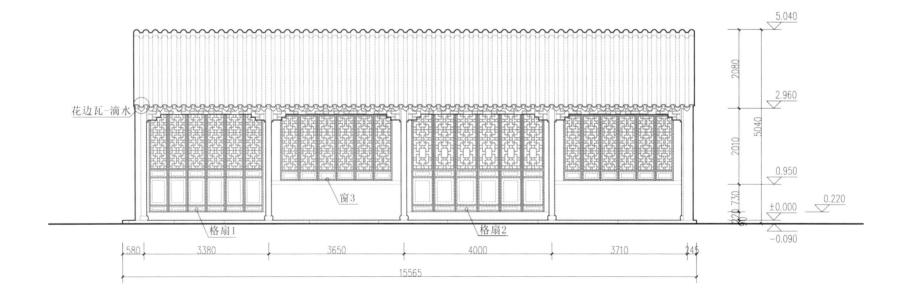

藕花水榭南立面图
South elevation: Ouhua Shuixie

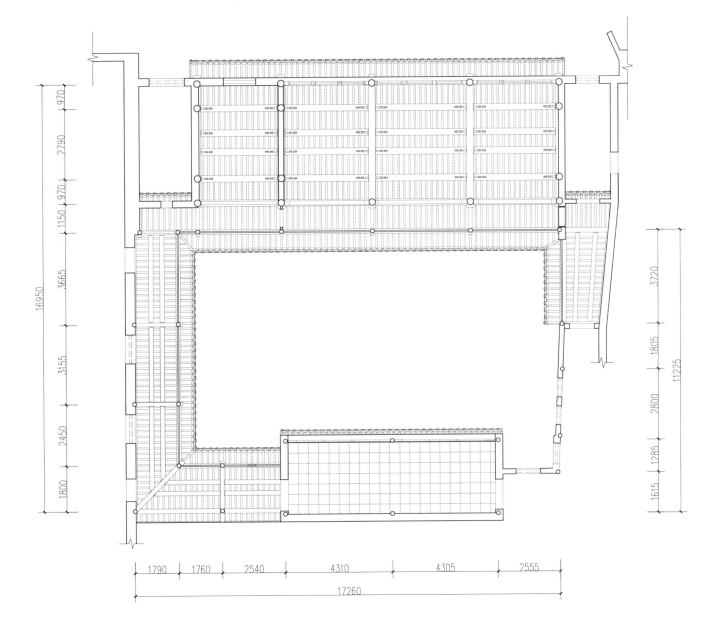

藕花水榭梁架仰视图
Reflected truss plan: Ouhua Shuixie

藕花水榭北面
Ouhua Shuixie: north side

藕花水榭院落
Ouhua Shuixie: courtyard

藕花水榭

Ouhua Shuixie

藕花水榭及办公室沿河北立面图

North elevation: Ouhua Shuixie and offices (riverside)

面水轩南面
Mianshui Xuan: South side

面水轩北面
Mianshui Xuan: North side

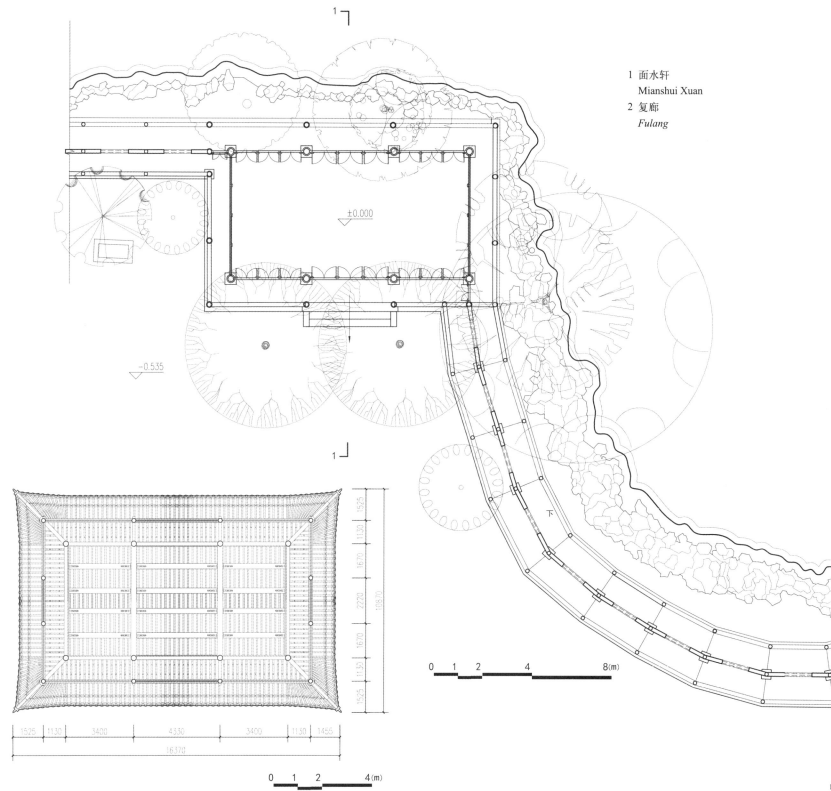

1 面水轩 Mianshui Xuan
2 复廊 Fulang

面水轩梁架仰视图
Reflected truss plan: Mainshui Xuan

面水轩区域平面图
Plan: Mainshui Xuan

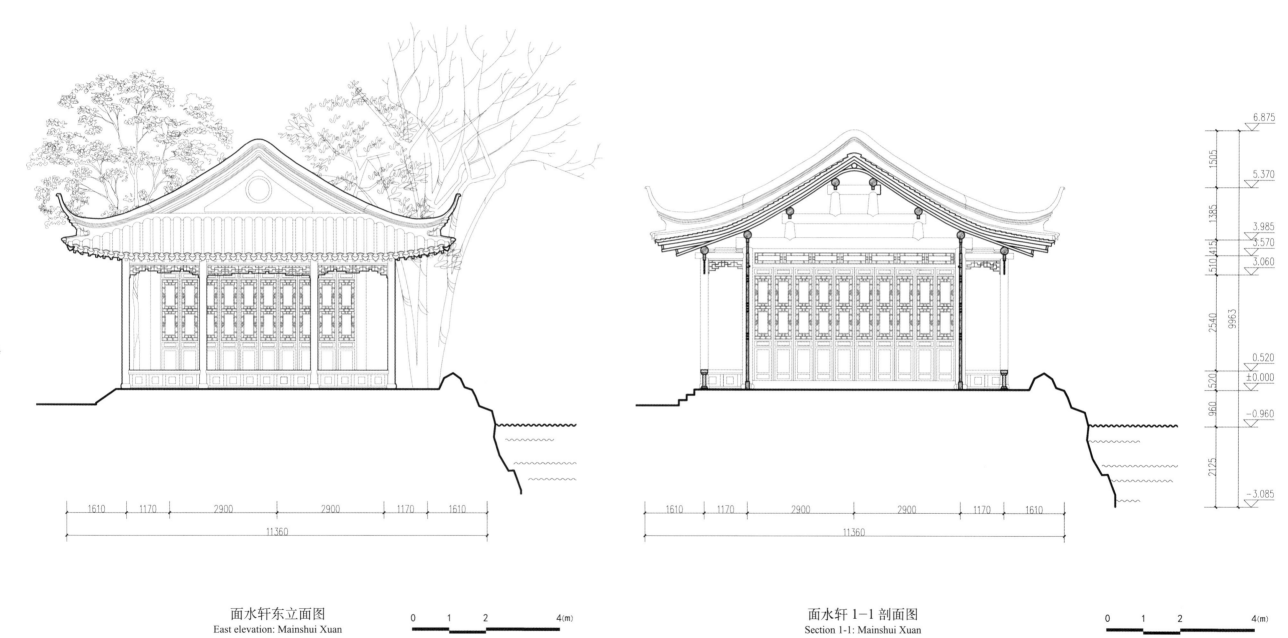

面水轩东立面图
East elevation: Mainshui Xuan

面水轩 1-1 剖面图
Section 1-1: Mainshui Xuan

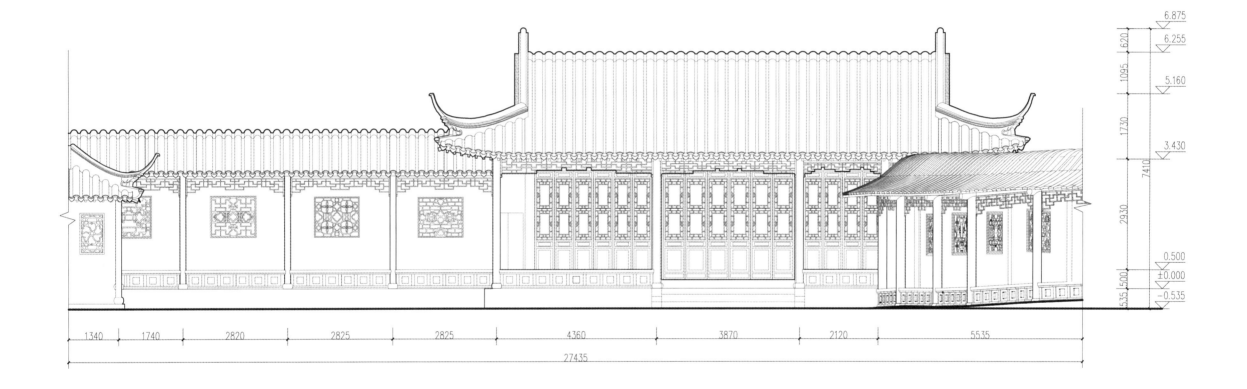

面水轩及东西廊南立面图

South elevation: Mianshui Xuan and galleries

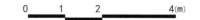

观鱼处
Guanyu Chu

面水轩及东西廊北立面图
North elevation: Mianshui Xuan and galleries

面水轩

Mianshui Xuan

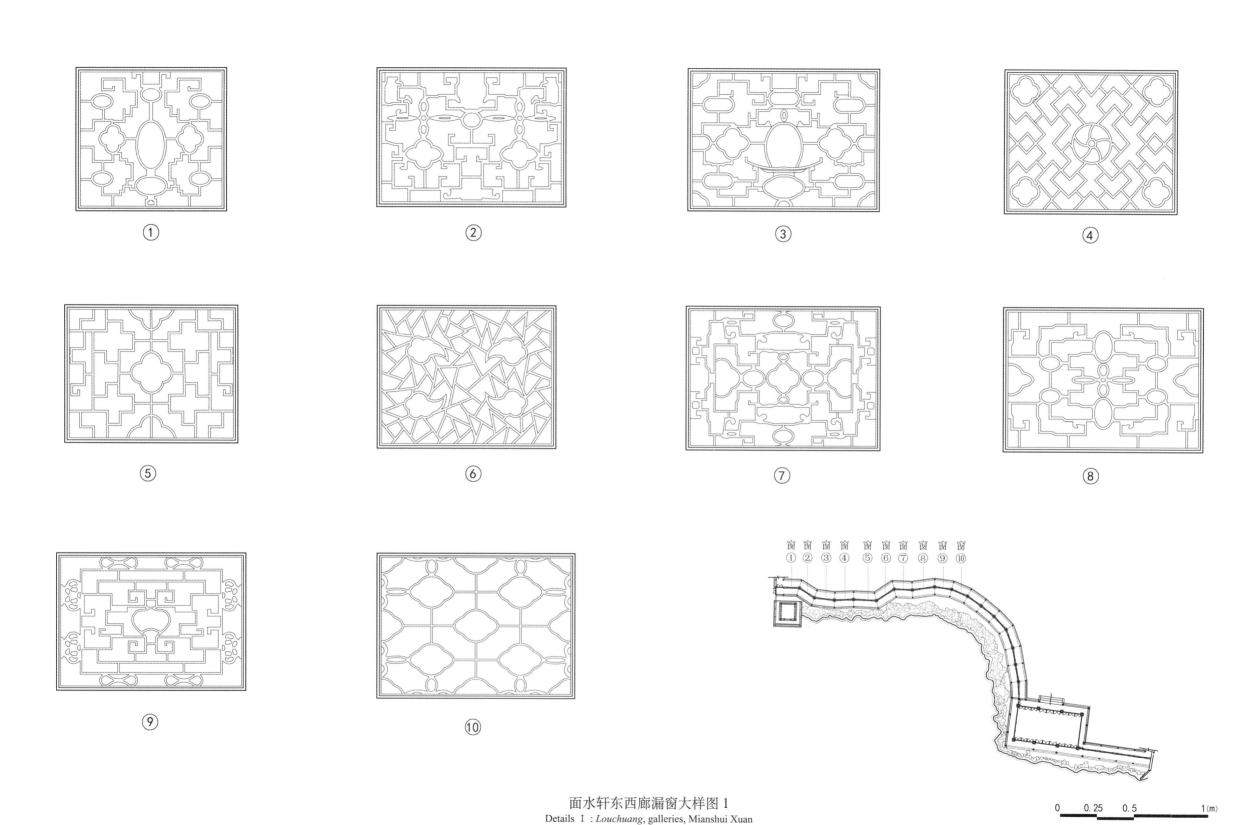

面水轩东西廊漏窗大样图1

Details Ⅰ: *Louchuang*, galleries, Mianshui Xuan

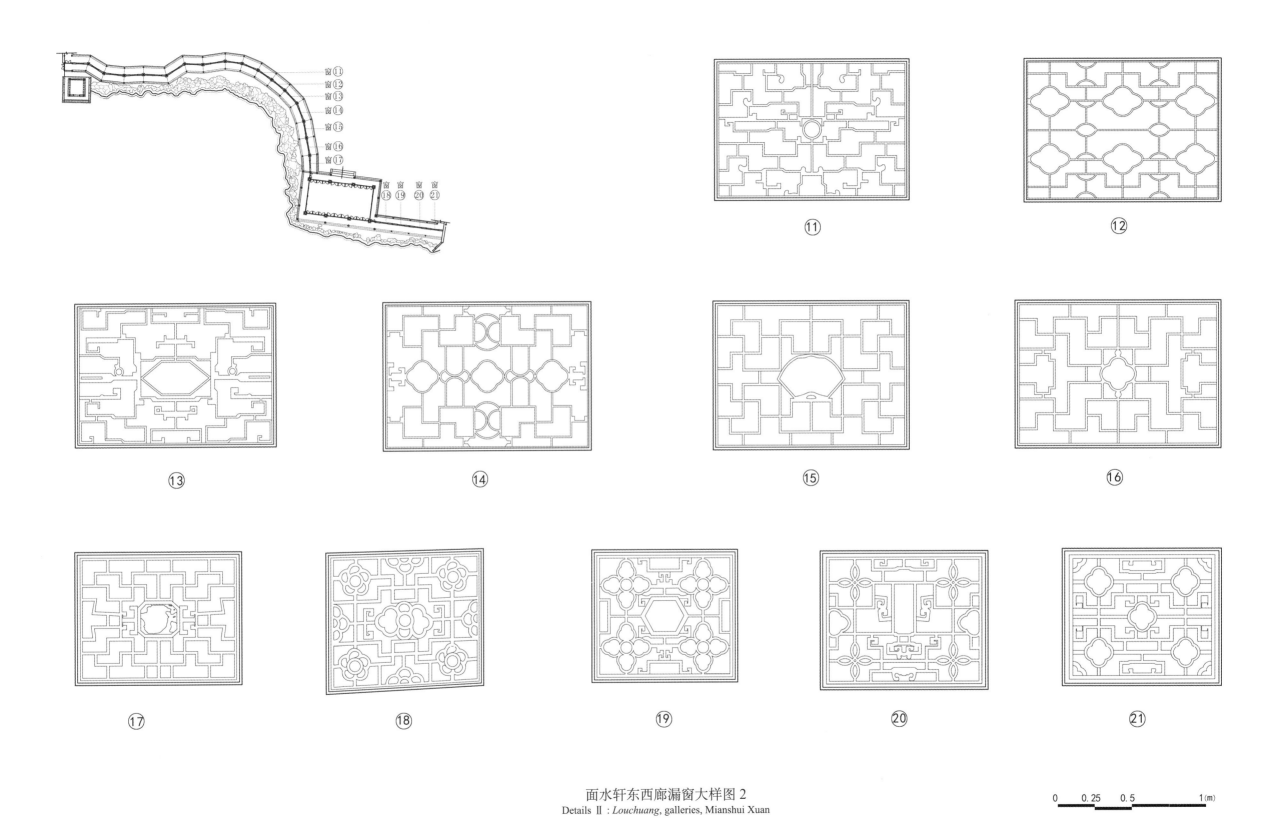

面水轩东西廊漏窗大样图 2
Details Ⅱ : *Louchuang*, galleries, Mianshui Xuan

观鱼处南侧
Guanyu Chu: south side

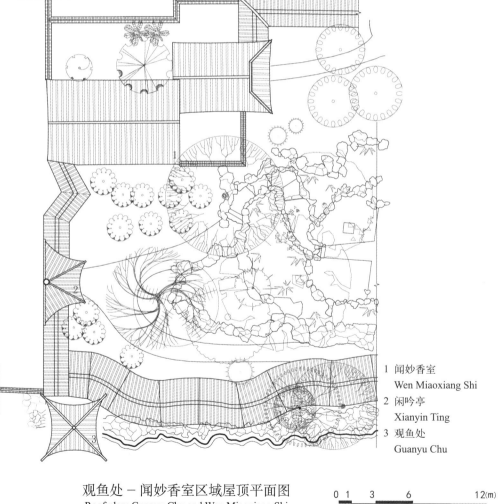

观鱼处-闻妙香室区域屋顶平面图
Roof plan: Guanyu Chu and Wen Miaoxiang Shi

1 闻妙香室 Wen Miaoxiang Shi
2 闲吟亭 Xianyin Ting
3 观鱼处 Guanyu Chu

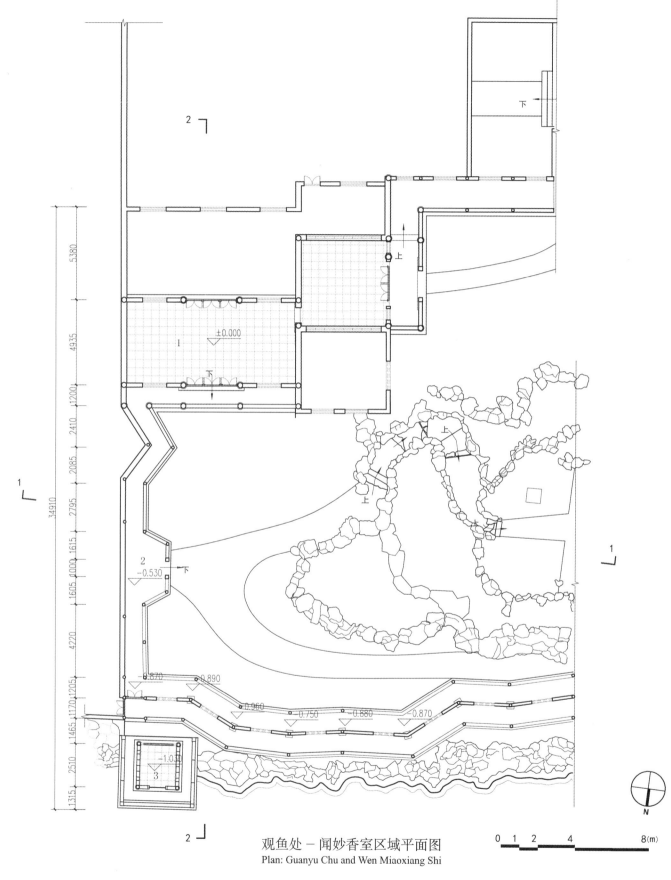

观鱼处-闻妙香室区域平面图
Plan: Guanyu Chu and Wen Miaoxiang Shi

观鱼处－闻妙香室区域 1-1 剖面图
Section 1-1: Guanyu Chu and Wen Miaoxiang Shi

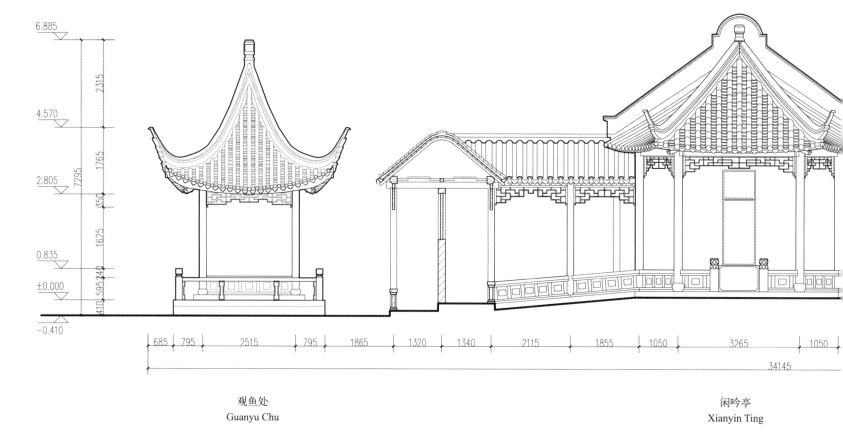

观鱼处 – 闻妙香室区域 2-2 剖面图
Section 2-2: Guanyu Chu and Wen Miaoxiang Shi

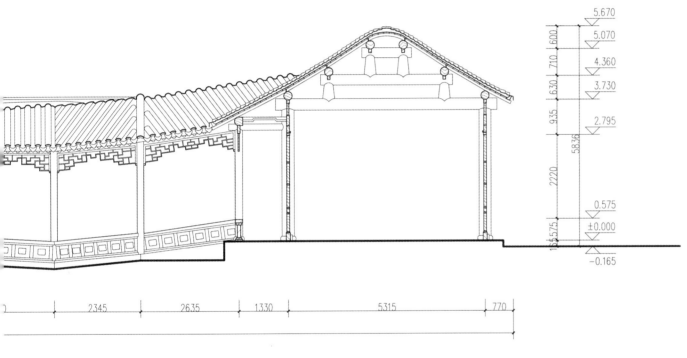

闻妙香室
Wen Miaoxiang Shi

真山林中部观沧浪亭
A view of Canglang Ting amid Zhen Shanlin area

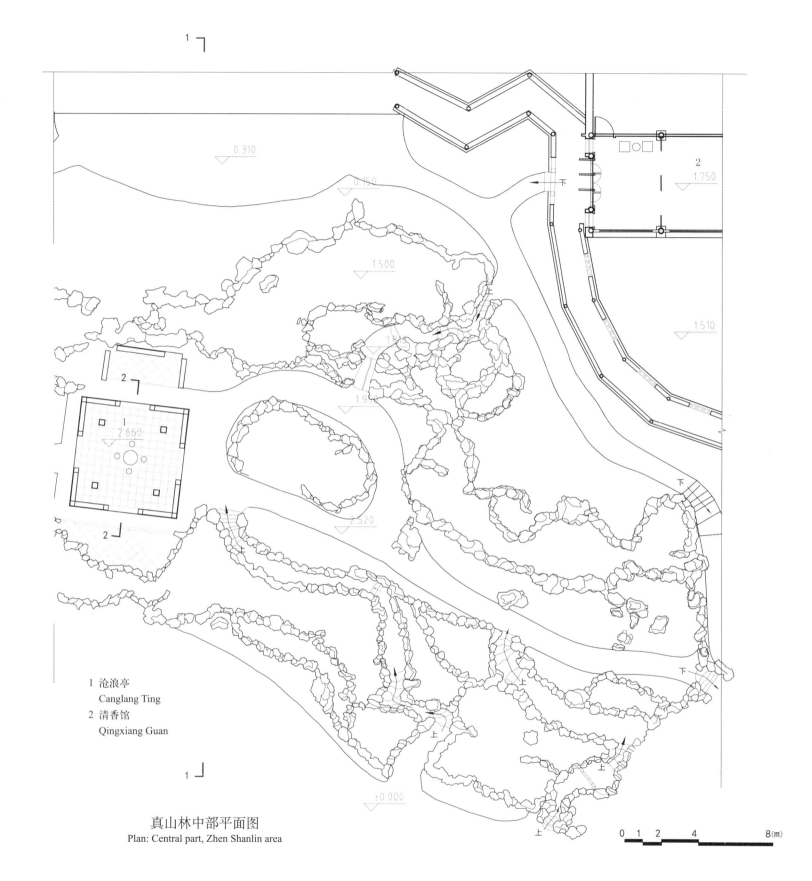

1 沧浪亭
　Canglang Ting
2 清香馆
　Qingxiang Guan

真山林中部平面图
Plan: Central part, Zhen Shanlin area

真山林中部 1-1 剖面图
Section 1-1: Central part, Zhen Shanlin area

复廊 *Fulang*
沧浪亭 Canglang Ting
闻妙香室 Wen Miaoxiang Shi

沧浪亭
Canglang Ting

真山林中部北立面图
North elevation: Central part, Zhen Shanlin area

沧浪亭
Canglang Ting

真山林中部南立面图
South elevation: Central part, Zhen Shanlin area

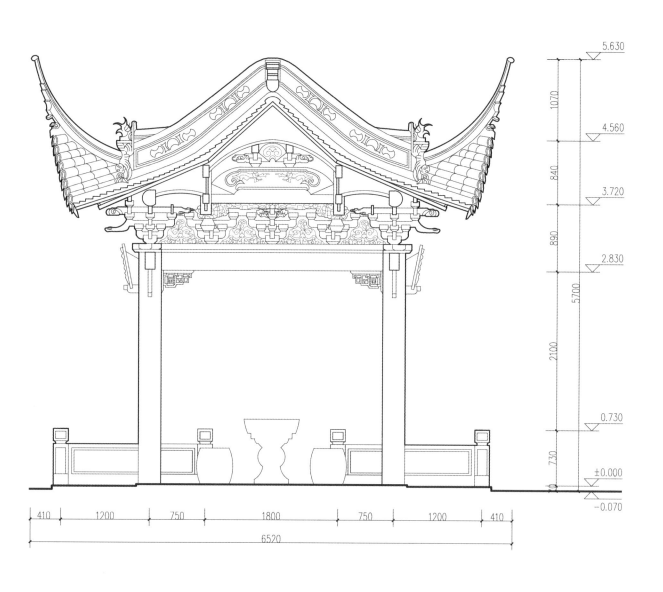

沧浪亭 2-2 剖面图
Section 2-2: Canglang Ting

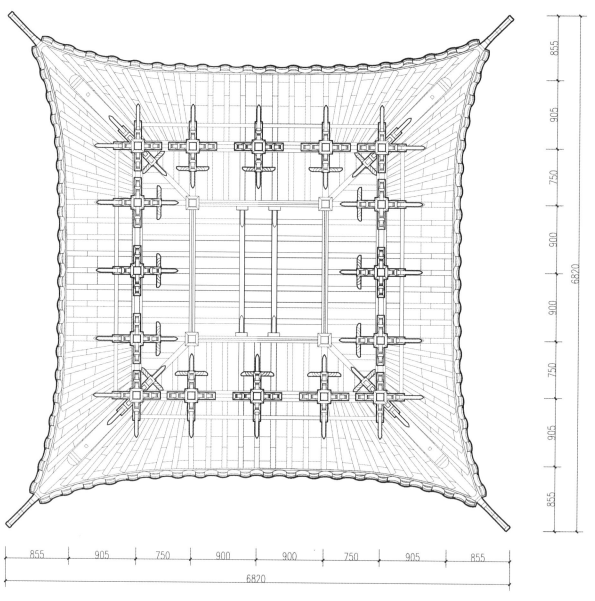

沧浪亭梁架仰视图
Reflected truss plan: Canglang Ting

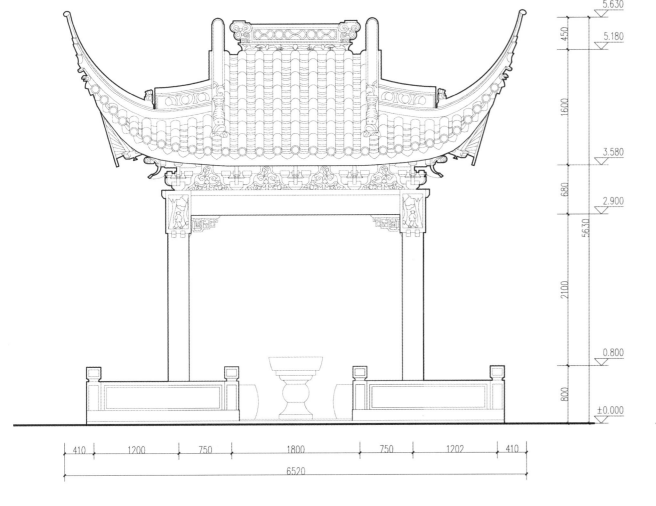

沧浪亭北立面图
North elevation: Canglang Ting

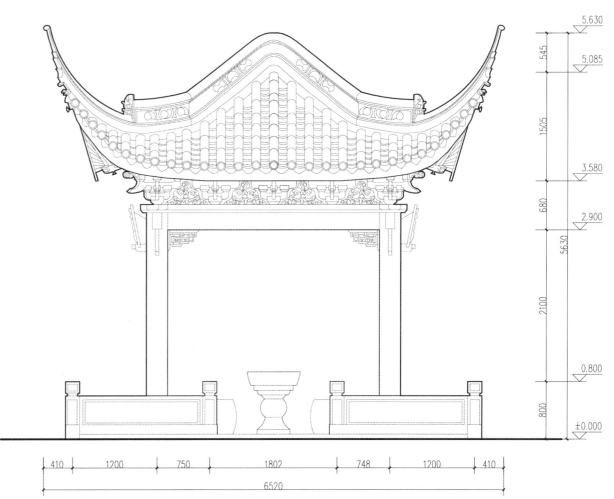

沧浪亭西立面图
West elevation: Canglang Ting

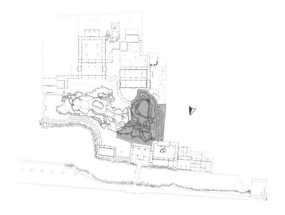

真山林西部
Zhen Shanlin: Western part

御碑亭
Yubei Ting

1 流玉潭
　Liuyu Tan
2 御碑亭
　Yubei Ting
3 步碕廊
　Buqi Lang

真山林西部－御碑亭区域平面图
Plan: Western part of Zhen Shanlin and Yubei Ting area

真山林西部区域 1-1 剖面图
Section 1-1: Western part, Zhen Shanlin area

     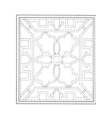 

漏窗 1–13 大样图
Details: *Louchuang* nos. 1–13

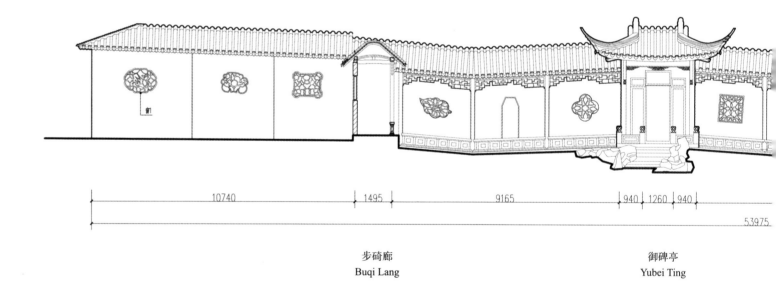

步碕廊
Buqi Lang

御碑亭
Yubei Ting

真山林西部 – 御碑亭区域 2–2 剖面图
Section 2-2: Western part, Zhen Shanlin and Yubei Ting area

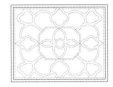

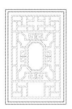

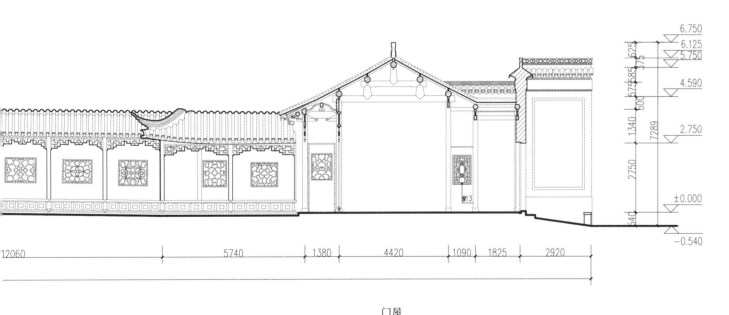

门屋
Gatehouse

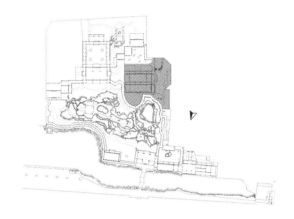

五百名贤祠院落
Wubai Mingxian Ci: courtyard

清香馆建筑室内
Qingxiang Guan: indoors

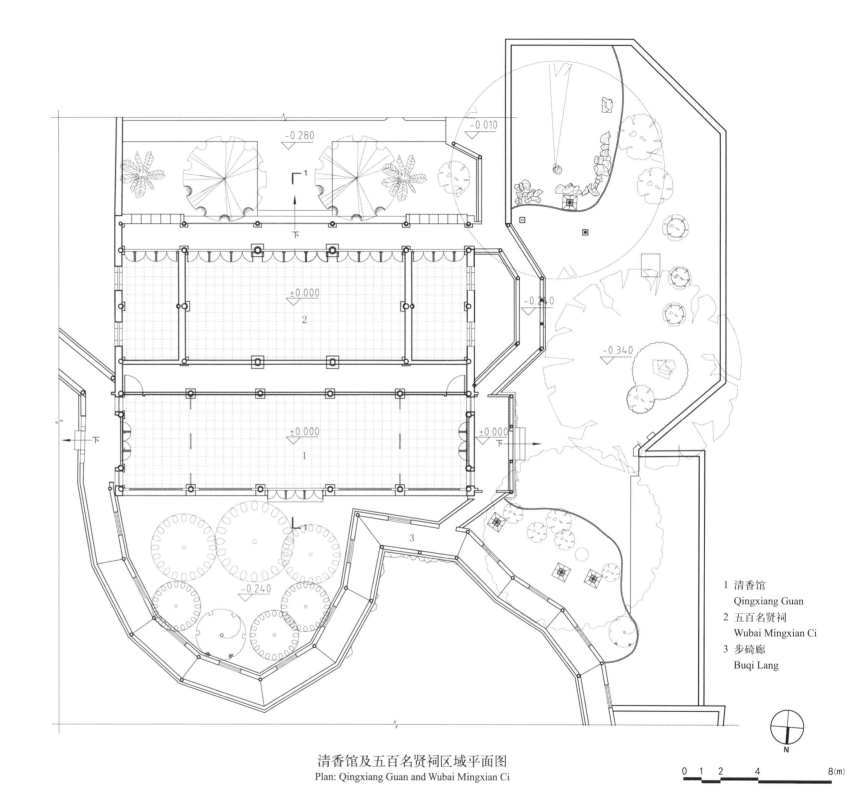

1 清香馆 Qingxiang Guan
2 五百名贤祠 Wubai Mingxian Ci
3 步碕廊 Buqi Lang

清香馆及五百名贤祠区域平面图
Plan: Qingxiang Guan and Wubai Mingxian Ci

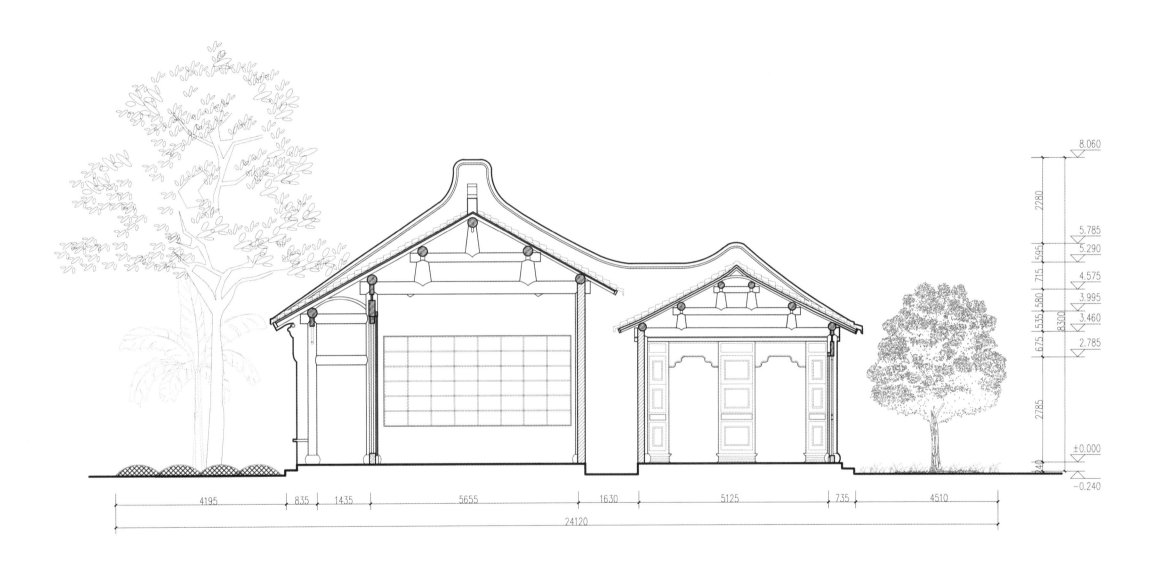

清香馆及五百名贤祠区域 1-1 剖面图
Section 1-1: Qingxiang Guan and Wubai Mingxian Ci

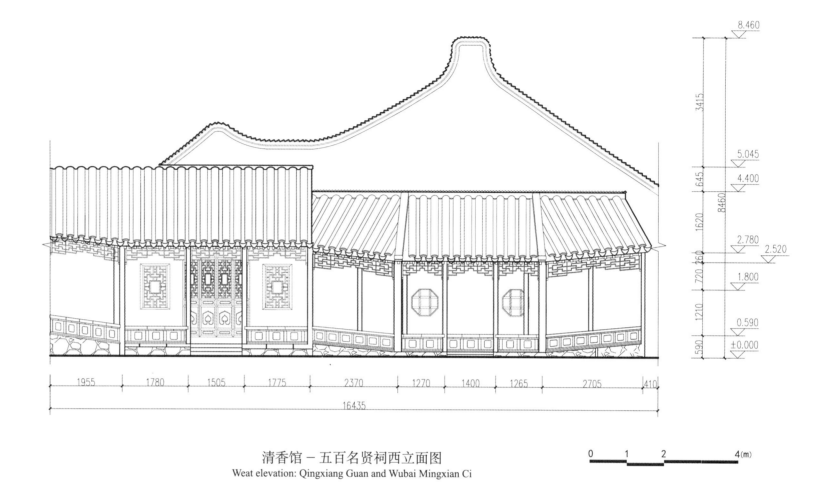
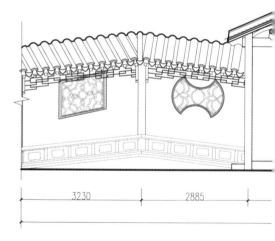

清香馆－五百名贤祠西立面图
Weat elevation: Qingxiang Guan and Wubai Mingxian Ci

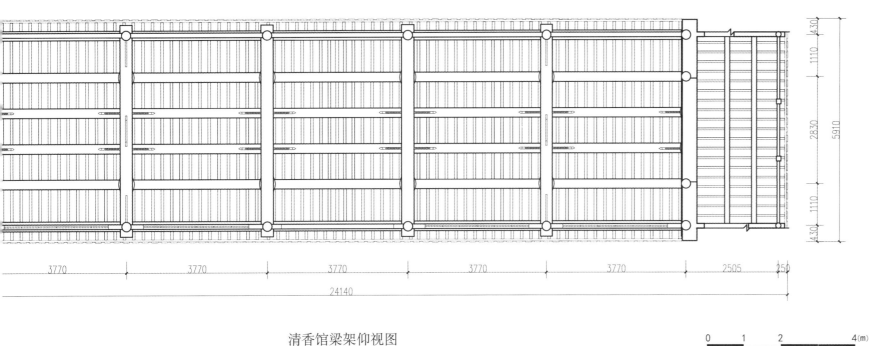

清香馆梁架仰视图
Reflected truss plan: Qingxiang Guan

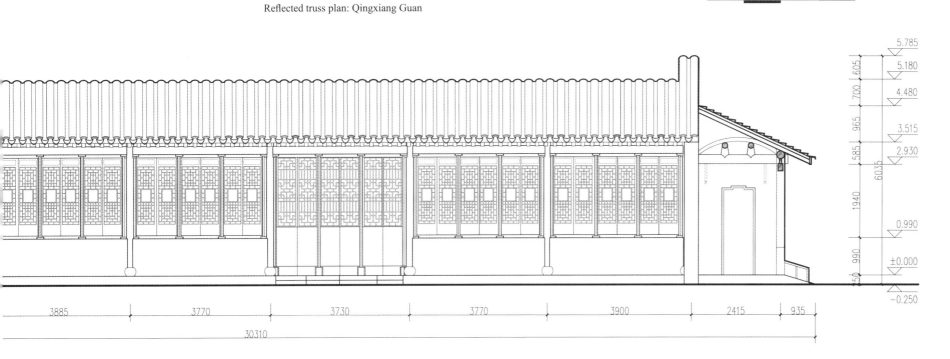

清香馆北立面图
North elevation: Qingxiang Guan

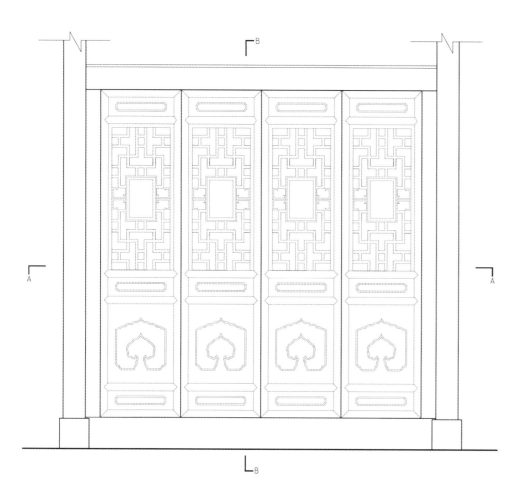
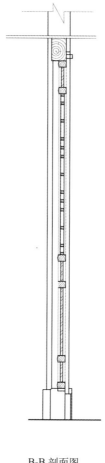
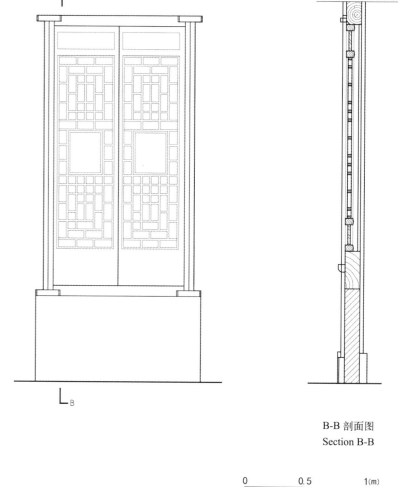

B-B 剖面图
Section B-B

B-B 剖面图
Section B-B

A-A 剖面图
Section A-A

清香馆隔扇大样图
Details: *Geshan*, Qingxiang Guan

清香馆槛窗大样图
Details: *Kanchuang* (partition window), Qingxiang Guan

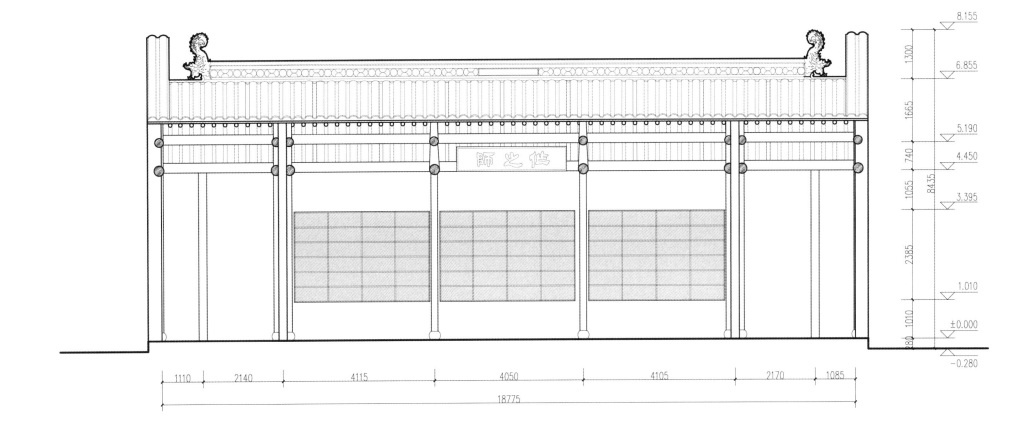

五百名贤祠纵剖面图
Longitudinal section: Wubai Mingxian Ci

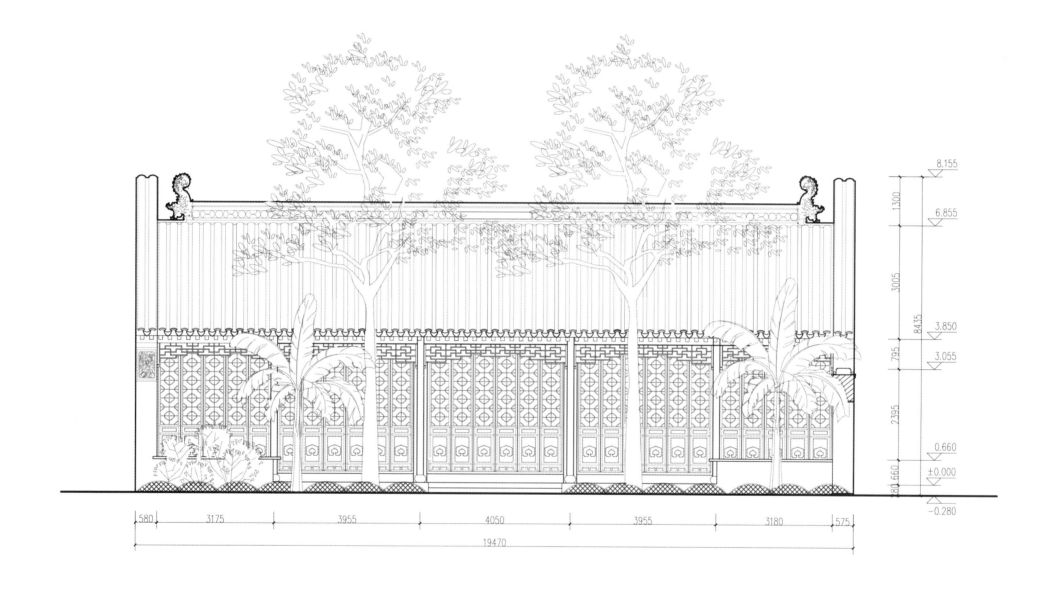

五百名贤祠南立面图
South elevation: Wubai Mingxian Ci

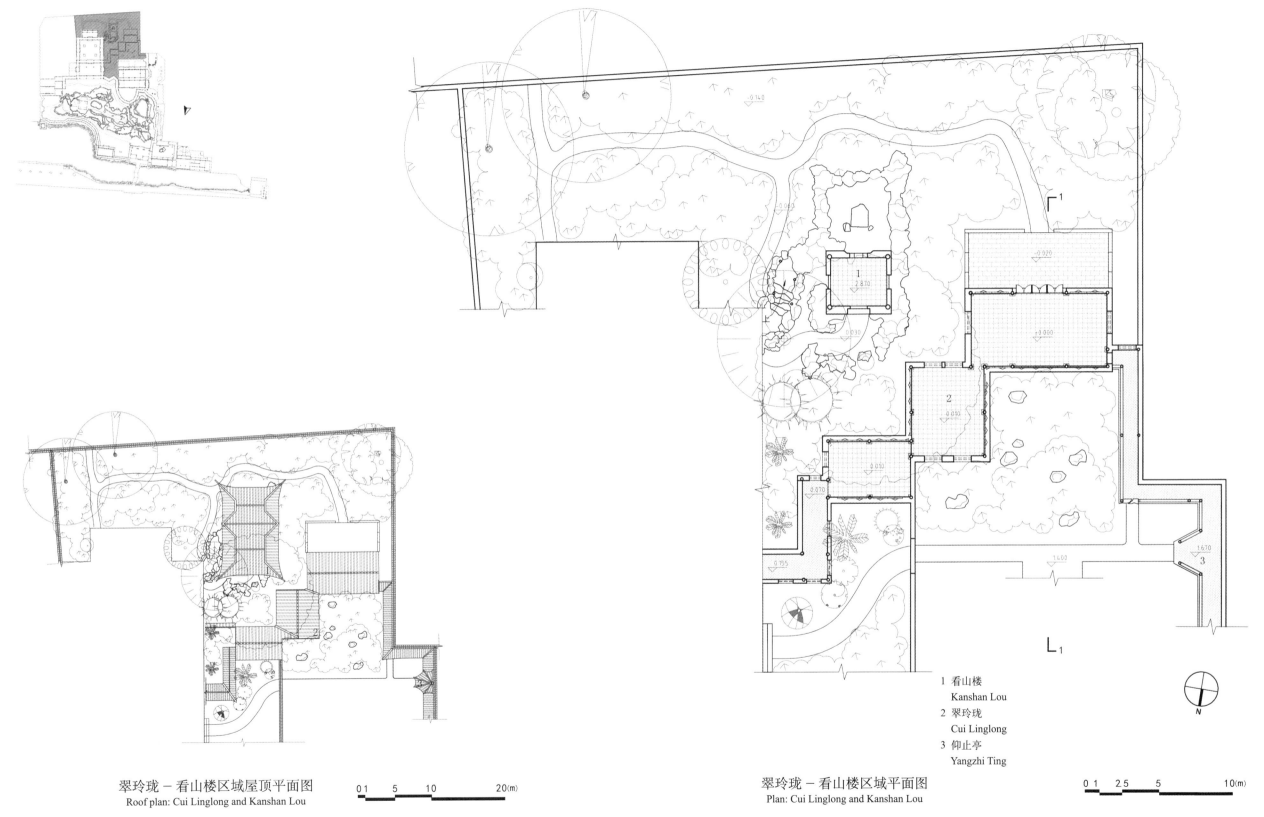

1 看山楼 Kanshan Lou
2 翠玲珑 Cui Linglong
3 仰止亭 Yangzhi Ting

翠玲珑－看山楼区域屋顶平面图
Roof plan: Cui Linglong and Kanshan Lou

翠玲珑－看山楼区域平面图
Plan: Cui Linglong and Kanshan Lou

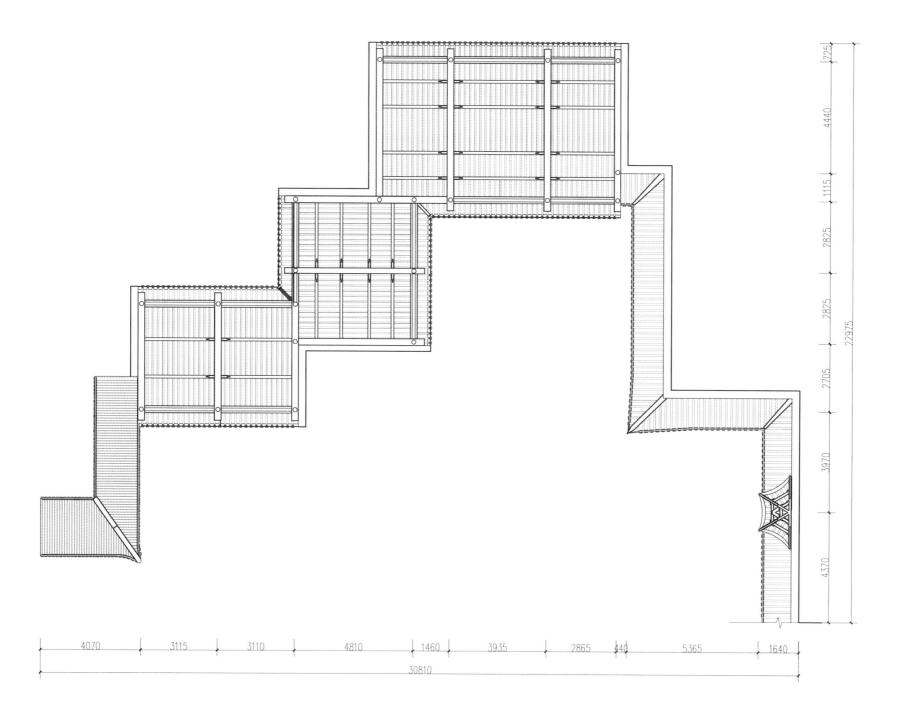

翠玲珑－仰止亭梁架仰视图

Reflected truss plan: Cui Linglong and Yangzhi Ting

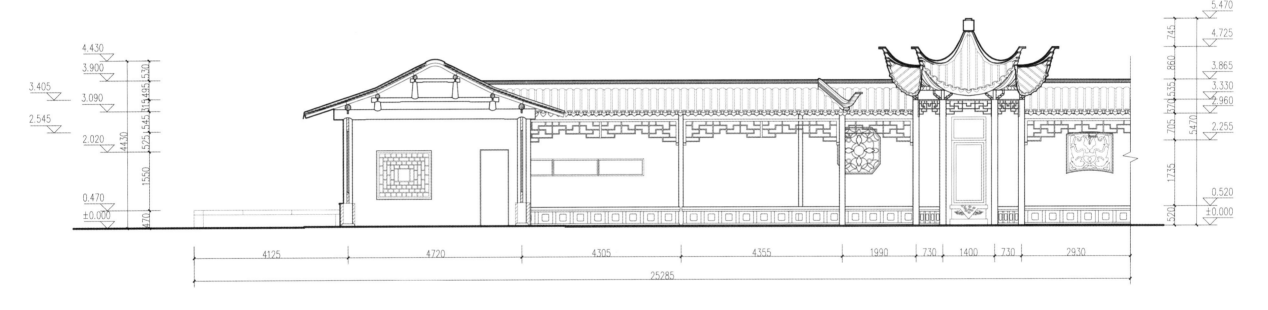

翠玲珑 – 仰止亭区域 1-1 剖面图
Section 1-1: Cui Linglong and Yangzhi Ting area

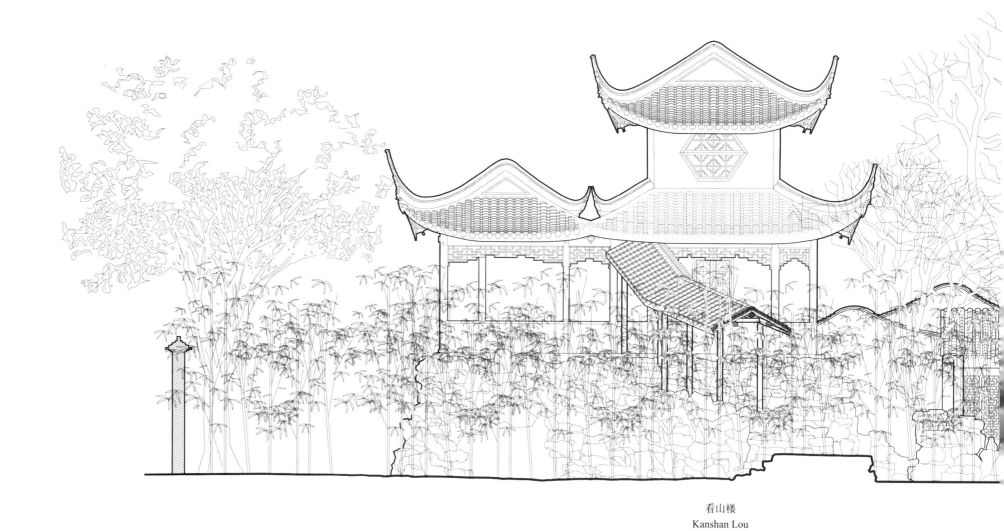

看山楼
Kanshan Lou

翠玲珑 – 看山楼区域东立面图
East elevation: Cui Linglong and Kanshan Lou

翠玲珑
Cui Linglong

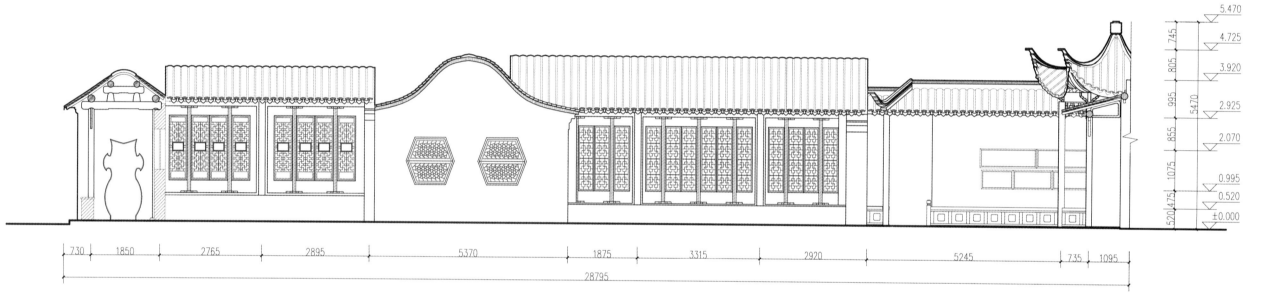

翠玲珑-仰止亭区域北立面图

North elevation: Cui Linglong and Yangzhi Ting

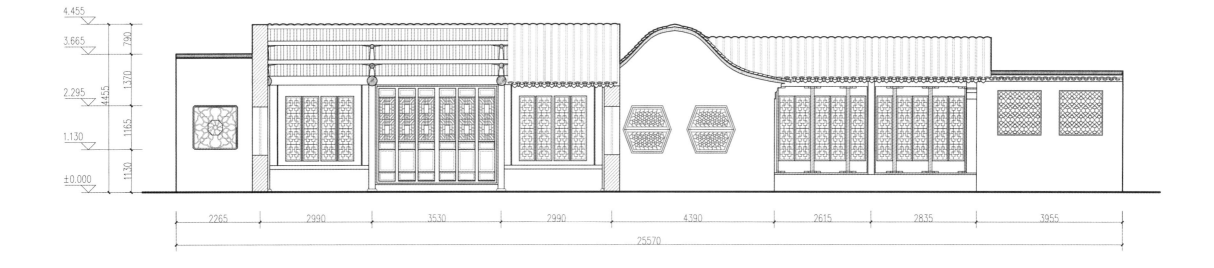

翠玲珑南剖立面图
South elevation: Cui Linglong

看山楼
Kanshan Lou

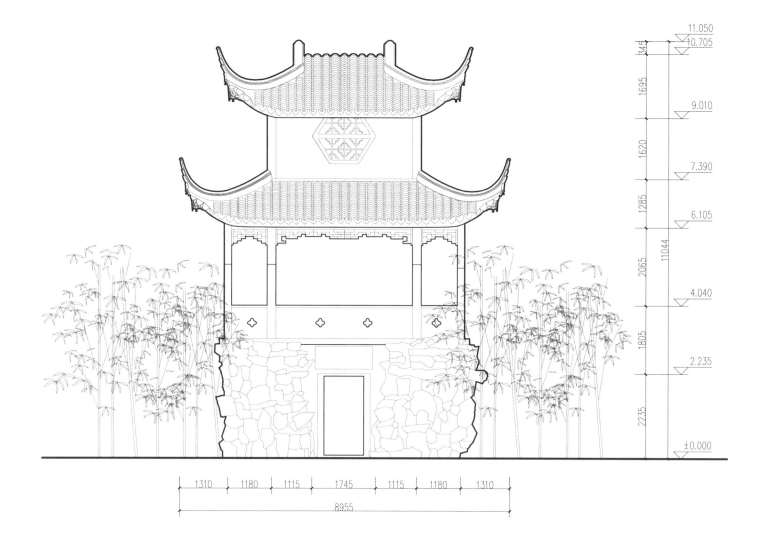

看山楼－印心石屋建筑北立面图
North elevation: Kanshan Lou and Yinxin Shiwu

1 瑶华境界
  Yaohua Jingjie
2 明道堂
  Mingdao Tang

明道堂
Mingdao Tang

瑶华境界
Yaohua Jingjie

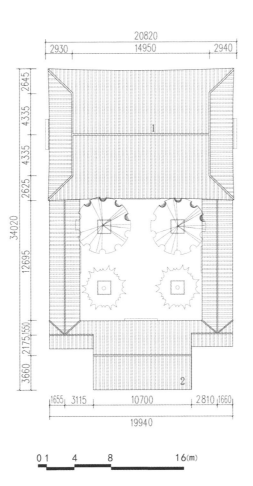

明道堂及瑶华境界区域屋顶平面图
Roof plan: Mingdao Tang and Yaohua Jingjie

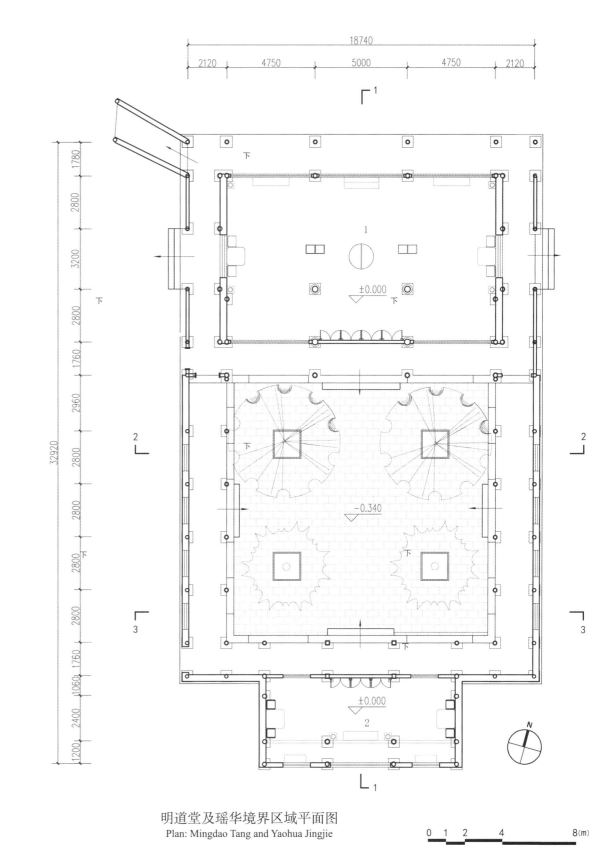

明道堂及瑶华境界区域平面图
Plan: Mingdao Tang and Yaohua Jingjie

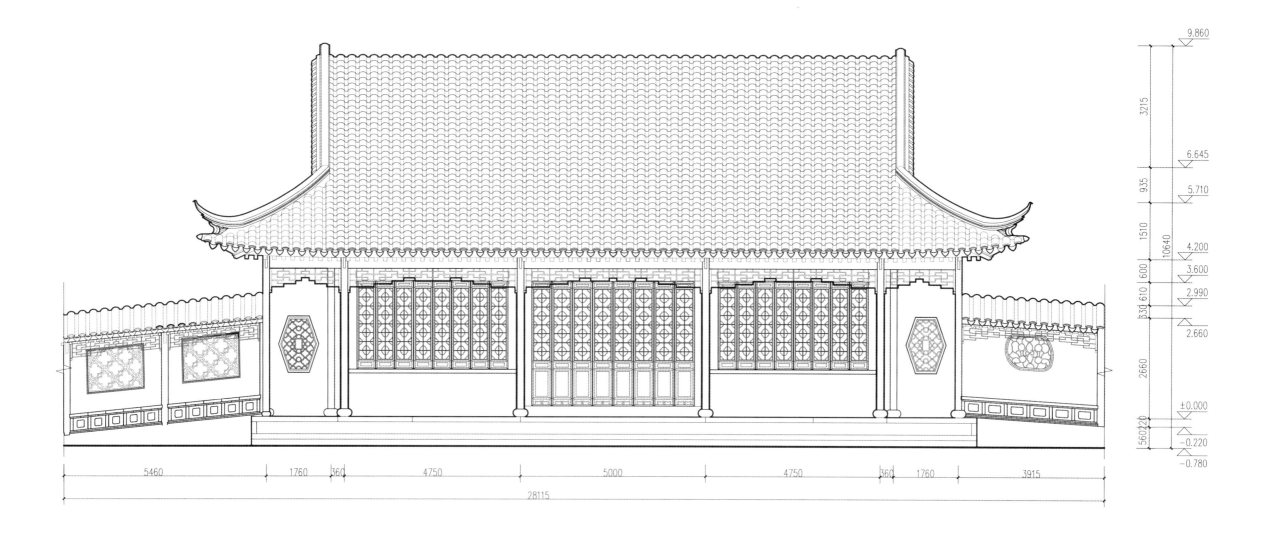

明道堂北立面图
North elevation: Mingdao Tang

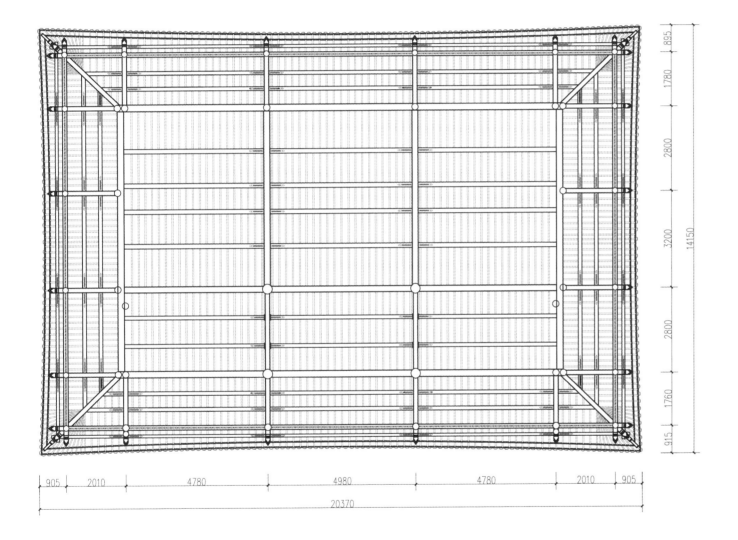

明道堂梁架仰视图
Reflected truss plan: Mingdao Tang

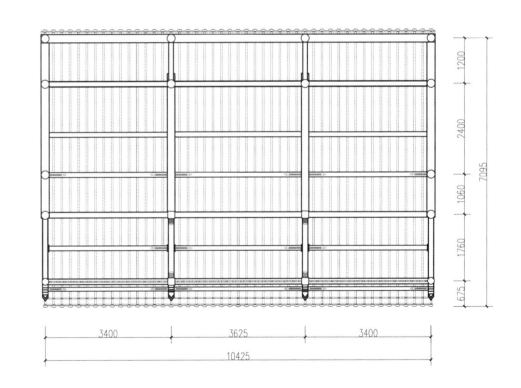

瑶华境界梁架仰视图
Reflected truss plan: Yaohua Jingjie

明道堂－瑶华境界院落模型
Model: Mingdao Tang and Yaohua Jingjie

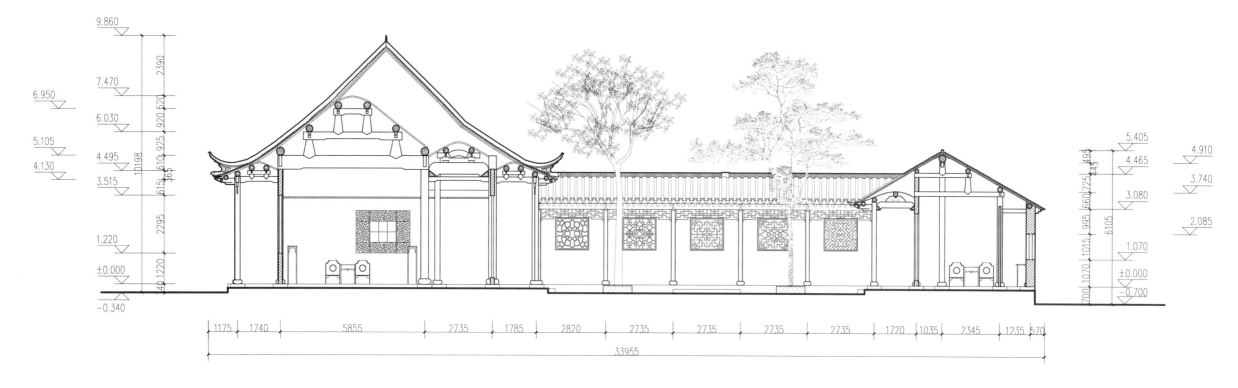

明道堂－瑶华境界院落 1-1 剖面图
Section 1-1: Mingdao Tang and Yaohua Jingjie

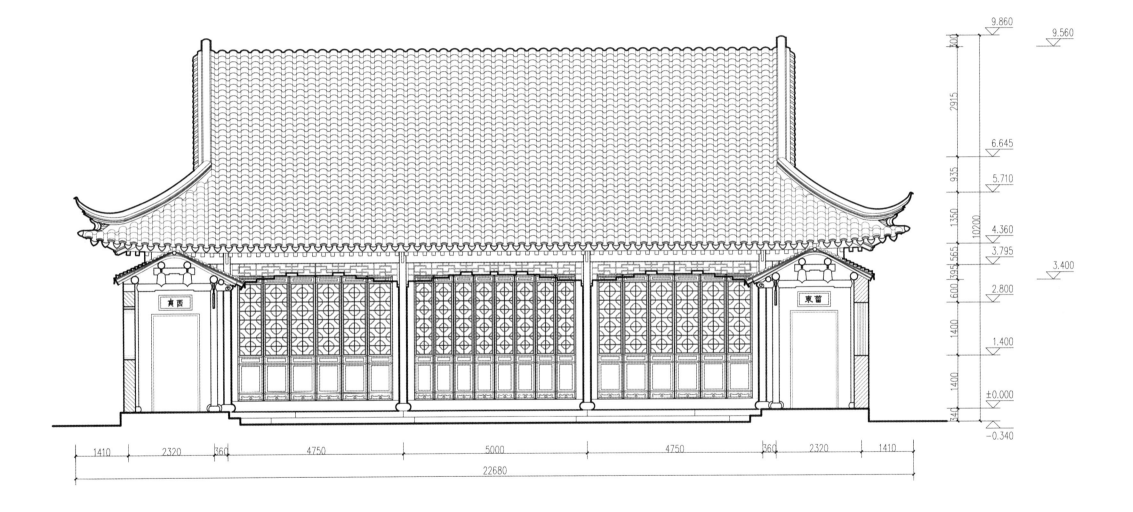

明道堂－瑶华境界院落 2-2 剖面图
Section 2-2: Mingdao Tang and Yaohua Jingjie

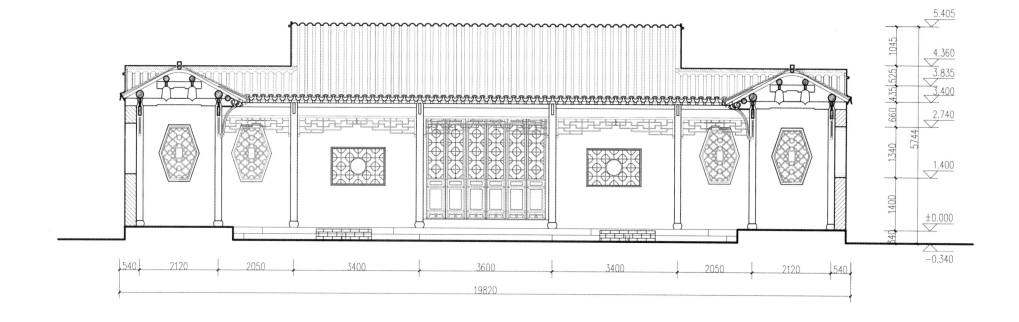

明道堂－瑶华境界院落 3-3 剖面图

Section 3-3: Mingdao Tang and Yaohua Jingjie

图 1

图 2

图 3

图 4

图 5

图 1　苏州艺圃
　　　Yi Pu, Suzhou

图 2　苏州环秀山庄
　　　Huanxiu Shanzhuang, Suzhou

图 3　苏州耦园
　　　Ou Yuan, Suzhou

图 4　苏州怡园
　　　Yi Yuan, Suzhou

图 5　苏州沧浪亭
　　　Canglang Ting, Suzhou

测绘工作记录
Surveyors in Working

后　记

本书收录的是我们在 2011—2014 年间测绘的五处苏州古典园林——艺圃、环秀山庄、怡园、耦园与沧浪亭的测绘图纸及相关成果。

2011 年，我首次开设了西安建筑科技大学的"古典园林测绘"课程，旨在拓展和丰富已有的"古建筑测绘"课程的教学内容。这对于我个人和我们教学团队而言均是测绘教学的新尝试。至 2021 年为止，十一年间我们指导建筑学、风景园林学和历史建筑保护工程专业的本科同学以及建筑历史方向的研究生们测绘了十处江南古典园林，除本书收入的五处之外，尚有苏州曲园、苏州网师园、苏州留园、无锡寄畅园以及扬州何园。同时我们也将测绘教学与古典园林的专题研究结合起来，现已有了初步成果，如《苏州艺圃》作为我们"一园一书"系列丛书的第一部，已于 2017 年由中国建筑工业出版社出版。

本书收录的五座园林，依据测绘外业开展的时间顺序排布：2011 年是艺圃，2012 年是环秀山庄，2013 年是耦园和怡园，2014 年则是沧浪亭。为了尽量呈现每座园子的现状特色与神韵风貌，并符合图书出版的相关要求，我们对图纸进行了全面整理。这些图纸为历年所集，卷帙浩繁，整理工作也艰巨繁冗，前后历时达两年之久，我在此感谢各位老师与研究生同学们在图纸绘制与整理工作中的全力投入！

感谢本书的责任编辑刘川女士和《建筑师》杂志主编李鸽博士，使本书得以付梓。

感谢大家的共同努力！

2022 年 9 月 20 日

Postscript

This volume presents to you the measured drawings, along with other findings, we made from 2011 to 2014 for the five classical gardens in Suzhou, namely Yi Pu, Huanxiu Shanzhuang, Yi Yuan, Ou Yuan and Canglang Ting.

Back in 2011, I launched *Classical Gardens Metric Survey,* a practical course aimed as a complement to the already established module, *Architectural Heritage Metric Survey,* for Xi'an University of Architecture and Technology (XAUAT). This meant a voyage into uncharted waters both for me and my team in terms of survey training. In the next 11 years until 2021, undergraduate architects, landscapers and architectural conservers as well as postgraduate architectural historians, under our tutelage, translated into drawings as many as ten classical gardens including Qu Yuan, Wangshi Yuan, Liu Yuan in Suzhou, Jichang Yuan in Wuxi as well as He Yuan in Yangzhou, plus the abovementioned five. These programs were coupled with follow-up studies on the gardens, which yielded various publications. Among them is *Yi Pu in Suzhou*, the first volume of our *One Book, One Garden* series, published in 2017 by China Building Industry Press.

Drawings included in this book are arranged chronologically as regards the surveys carried out, that is, Yi Pu (2011), Huanxiu Shanzhuang (2012), Ou Yuan (2013), Yi Yuan (2013) and Canglang Ting (2014). To render as fully as possible the status, appearance and charm of these gardens, as well as to comply with certain publishing regulations, meticulous selection and preparation of the drawings were conducted. Considering the myriad drawings generated by our extensive surveys, the accomplishment of this back-breaking and time-consuming, 2 years, job should be ascribed to every teaching staff member and postgraduates in our team.

Special thanks should go to LIU Chuan, the editor-in-charge of the book and LI Ge, the chief editor of *The Architect*, the journal, who handled the publishing of this book.

I hereby thank sincerely everybody whose joint effort made happen this publication.

LIN Yuan
20th September, 2022

图书在版编目（CIP）数据

尘外画中：西安建筑科技大学古典园林测绘图辑：2011—2014 = Otherworldliness Depicted: Measured Drawings for Classical Gardens by Xi'an University of Architecture and Technology, 2011—2014：汉英对照 / 林源等著 . —北京：中国建筑工业出版社，2019.12

ISBN 978-7-112-24597-0

Ⅰ.①尘… Ⅱ.①林… Ⅲ.①古典园林—建筑测量—苏州—图集 Ⅳ.① TU986.62-64

中国版本图书馆CIP数据核字（2020）第018393号

责任编辑 / 李鸽　刘川
责任校对 / 王烨

尘外画中
—— 西安建筑科技大学古典园林测绘图辑2011-2014

林源　岳岩敏　汶武娟　林溪　著

Otherworldliness Depicted:
Measured Drawings for Classical Gardens by Xi'an University of Architecture and Technology, 2011-2014
LIN Yuan, YUE Yanmin, WEN Wujuan, LIN Xi

*

中国建筑工业出版社出版、发行（北京海淀三里河路9号）
各地新华书店、建筑书店经销
北京方舟正佳图文设计有限公司制版
北京富诚彩色印刷有限公司印刷

*

开本：787毫米×1092毫米　横1/8　印张：33½　字数：821千字
2022年11月第一版　　2022年11月第一次印刷
定价：**258.00元**
ISBN 978-7-112-24597-0
　　　（35220）

版权所有　翻印必究
如有印装质量问题，可寄本社图书出版中心退换
（邮政编码100037）